Springer Theses

Recognizing Outstanding Ph.D. Research

Aims and Scope

The series "Springer Theses" brings together a selection of the very best Ph.D. theses from around the world and across the physical sciences. Nominated and endorsed by two recognized specialists, each published volume has been selected for its scientific excellence and the high impact of its contents for the pertinent field of research. For greater accessibility to non-specialists, the published versions include an extended introduction, as well as a foreword by the student's supervisor explaining the special relevance of the work for the field. As a whole, the series will provide a valuable resource both for newcomers to the research fields described, and for other scientists seeking detailed background information on special questions. Finally, it provides an accredited documentation of the valuable contributions made by today's younger generation of scientists.

Theses are accepted into the series by invited nomination only and must fulfill all of the following criteria

- They must be written in good English.
- The topic should fall within the confines of Chemistry, Physics, Earth Sciences, Engineering and related interdisciplinary fields such as Materials, Nanoscience, Chemical Engineering, Complex Systems and Biophysics.
- The work reported in the thesis must represent a significant scientific advance.
- If the thesis includes previously published material, permission to reproduce this must be gained from the respective copyright holder.
- They must have been examined and passed during the 12 months prior to nomination.
- Each thesis should include a foreword by the supervisor outlining the significance of its content.
- The theses should have a clearly defined structure including an introduction accessible to scientists not expert in that particular field.

More information about this series at http://www.springer.com/series/8790

Nina-Louisa Remuss

Theorising Institutional Change: The Impact of the European Integration Process on the Development of Space Activities in Europe

Doctoral Thesis accepted by
the University of Potsdam, Potsdam, Germany

Springer

Author
Dr. Nina-Louisa Remuss
Berlin, Germany

Supervisors
Prof. Kai-Uwe Schrogl
University of Tübingen
Tübingen, Germany

Prof. Falk Daviter
University of Potsdam
Potsdam, Germany

ISSN 2190-5053 ISSN 2190-5061 (electronic)
Springer Theses
ISBN 978-3-030-07121-9 ISBN 978-3-319-95978-8 (eBook)
https://doi.org/10.1007/978-3-319-95978-8

This Springer imprint is published by the registered company Springer Nature Switzerland AG
The registered company address is: Gewerbestrasse 11, 6330 Cham, Switzerland

Supervisor's Foreword

The Ph.D. thesis of Mrs. Nina-Louisa Remuss deals with the governance of European space activities, which has been in transition for some time, thereby generating a growing political and academic discourse. It is a topic of growing practical, academic and political importance on the European level. Central to this, is the question about the future role of the European Space Agency (ESA), which was founded in 1975 as a multilateral European organisation outside the European Communities/European Union. Today ESA carries out space activities for currently 22 Member States. ESA's role will depend on the EU's capacity to extend its competences to space policy, which is currently politically enforced.

This thesis thus deals with a current topic with great practical relevance, since space policy is of decisive importance for industrial policy and has numerous applications for the EU's policy areas. The existence of two European Organisations—the supranational EU and the multilateral ESA—is a growing political challenge, which is highlighted by the intensified dispute and institutional competition. Against this background, the topic of this dissertation is not only of great political relevance but also an exceptional subject for a political science analysis.

Institutional change can be observed in diverse ways, but the case at hand offers a great opportunity to analyse institutional change of two international organisations, working in the same policy field and being of different status (supranational and multilateral). Mrs. Remuss has chosen and worked on a particularly interesting topic, which is extraordinarily suitable to deliver interesting insights from a political science point of view.

Central to this work is the analyses of both the change in the development of space policy in Europe as well as the institutional change in both EU and ESA. This dissertation follows four objectives:

1. To provide an updated overview of the development of space policy in Europe
2. To advance the debate about the impact of the European integration process on existing institutional actors

3. To make a theoretical contribution to our understanding of when supranational actors are able to advance their own agenda
4. To make a theoretical contribution to our understanding of institutional change.

It is proposed to rely on a model of five stages of institutional development based on Levy, Young, & Zürn (1995) and Florensa (2004). Additionally, in order to be able to compare cases of institutional change, it is proposed to rely on five facets of institutional change. Both, the stages of institutional development and the facets of institutional change, are then applied to the development of space activities in Europe from 1959 to 2014.

The empirical content of this work is characterised by Mrs. Remuss' extraordinary expertise. Prior to working for the German Aerospace Center (DLR), Mrs. Remuss gained substantial insights while working for the European think tank European Space Policy Institute (ESPI). Her publications from this period guaranteed the concise presentation of the development of space activities in Europe, which goes beyond the existing literature. It offers new empirical findings for the presentation of the development of space activities in Europe and its institutions.

On this solid basis, this is a profound work, which claims to make a theoretical contribution to understanding institutional change. Particularly, the analysis of the impact of the European integration process on existing institutional actors and the case of the relations between EU and ESA are illustrated to an unprecedented extent.

Tübingen, Germany
June 2018

Prof. Dr. Kai-Uwe Schrogl

References

Florensa, M. C. (2004). *Institutional Stability and Change. A Logic Sequence for Studying Institutional Dynamics*. Paper presented at the Tenth Biennial Conference of the International Association for the Study of Common Property (IASCP). Oaxaca, Mexico. Retrieved from: https://dlc.dlib.indiana.edu/dlc/bitstream/handle/10535/1823/CostejaFlorensa_Institutional_040525_Paper220.pdf?sequence=1 [accessed 13 March 2017].

Levy, M. A., Young, O. R. & Zürn, M. (1995). The Study of International Regimes. *European Journal of International Relations, 1*, pp. 267–330.

Abstract

While European Space collaboration was initially developed outside the competences of the European Union (EU) with space programmes being carried out almost exclusively under the framework of European Space Agency (ESA) and national agencies, the EU has gained "shared competences" (Art. 2, TFEU) in space policy following the adoption of the Lisbon Treaty. Currently, the EU and ESA work together under a Framework Agreement. In 2016, the EU Commission has published a Communication entitled "European Space Policy" (ESP). Even though ESA's Member States have agreed to keep ESA as an intergovernmental organisation during the ESA Ministerial Council of 2014, the discussion about ESA becoming part of the EU framework continues. The EU's ambitions for leadership in European space policy raise question concerning the future of ESA. The study of institutions lies at the heart of political sciences. Strikingly, the theoretic framework qualifying institutional change and making it comparable leaves room for more concrete and testable dimensions of institutional change. This dissertation thus seeks to advance the debate on institutional change by proposing a more structured approach to institutional change, consisting of stages of institutional development and facets of institutional change. Thereby, the evolving institutional arrangements between EU and ESA are considered a case study. Special attention is paid to the impact of the European integration process on existing institutional actors.

Acknowledgements

I would like to thank my supervisor **Prof. Dr. Falk Daviter** for his valuable suggestions and criticism, which made this work as strong and precise as it is today.

Special thanks go to **Prof. Dr. Lesley Jane Smith** and **Prof. Dr. Kai-Uwe Schrogl,** whose support was incredibly helpful for my work. Without their insights and support this work would have been impossible.

Moreover, I would like to thank **Prof. Dr. Andrea Liese**, who complemented the aforementioned in supervising the defence of this piece.

I would also like to show my gratitude to my parents, **Marlies and Bodo Remuss**, and my boyfriend, **Frank Bowinkelmann**, who always believed in me and my skills. Without their continuous support this piece would have been impossible.

Finally, my gratitude goes to the city of **Warnemünde**, which provided not only the ideal location to focus but also sufficient motivation during a period of writing.

Contents

Abbreviations and Acronyms

AFC	Administrative and Finance Committee
ASI	Agencia Espaciale Italiana—Italian Space Agency
CCMI	Consultative Commission for Industrial Change
CEPT	Conférence Européenne des Administrations des Postes et des Télécommunications—European Conference of Postal and Telecommunications Administrations
CERN	European Organisation for Nuclear Research
CETS	European Conference on Satellite Communications
CFSP	Common Foreign and Security Policy
CNES	Centre National d'études Spatiales—French Space Agency
COM	European Commission/Commission of the European Communities
COPERS	European Preparatory Commission for Space Research
COSPAR	Committee of Space Research
CWG	Council Working Group
DG	Director General
DGs	Directorate-Generals
DLR	Deutsches Zentrum für Luft- und Raumfahrt—German Space Agency
DoD	Department of Defense
EC	European Community
EDA	European Defence Agency
EEC	European Economic Community
EESC	European Economic and Social Committee
ELDO	European Launcher Development Organisation
EO	Earth Observation
EP	European Parliament
ESA	European Space Agency
ESC	European Space Conference
ESDAC	European Data Analysis Centre
ESDP	European Security and Defence Policy

ESOC	European Space Operations Centre
ESP	European Space Policy
ESRIN	ESA Centre for Earth Observation
ESRO	European Space Research Organisation
ESS	European Space Strategy
ESTEC	European Space Technology Centre
EU	European Union
EURATOM	European Atomic Energy Community
FA	Framework Agreement
FRA	France
GEERS	Group d'Etudes Européen pour la Collaboration dans le Domaine des Recherches Spatiales or European Space Research Study Group
GER	Germany
GNSS	Global Navigation Satellite System
GPS	Global Positioning System
GSA	GNSS Supervisory Authority
IGC	Intergovernmental Conference
IRAC	International Relations Advisory Committee
ISS	International Space Station
JAA	Joint Aviation Authorities
JSSAG	Joint Space Strategy Advisory Group
JTF	Joint Task Force
LAFWG	Legal, Administrative and Financial Working Group
LI	Liberal Intergovernmentalism
LPAC	Launching Programme Advisory Committee
MoU	Memorandum of Understanding
MS	Member States
NATO	North Atlantic Treaty Organization
NEO	New Economics of Organisation
OEEC	Organization for European Economic Cooperation
OSCE	Organisation for Security and Co-operation in Europe
R&D	Research and Development
RB	Roland Berger
SAG	Space Advisory Group
SatCen	European Union Satellite Centre
SCG	Space Coordination Group
SEA	Single European Act
STC	Scientific and Technical Committee
STWG	Scientific and Technical Working Group
TCE	Treaty establishing a Constitution for Europe
TEC	Treaty establishing the European Community
TEU	Treaty on European Union
TFEU	Treaty on the Functioning of the European Union
UK	United Kingdom
U.S.	United States

USSR	Union of Soviet Socialist Republics
WEAG	Western European Armaments Group
WEAO	Western European Armaments Organisation
WEU	Western European Union
WG	Working Group

List of Figures

Chapter 1
Introduction

While European space collaboration was initially developed outside the competences of the European Union (EU) with space programmes being carried out almost exclusively under the framework of the European Space Agency (ESA) and national agencies, the EU has gained "shared competences" (Art. 2, TFEU) in space policy following the adoption of the Lisbon Treaty. The EU and ESA have already been working together under a Framework Agreement (FA) since 2003. Even though ESA's Member States have agreed to keep ESA as an intergovernmental organisation during the ESA Ministerial Council of 2014, the discussion about ESA becoming part of the EU framework continues. The EU's ambitions for leadership in European space policy raise questions concerning the governance of space in Europe and the inter-institutional relationship of the EU and ESA. While the institutional problem of setting-up a structure for a coherent European space policy has already been recognised in the early 90s (cf. Schrogl, 1993), the question of the institutional arrangement is still not solved today even though it has been the subject of much academic and political debate in recent years (cf. Hörber, 2016a; Gaubert, 2006).

Although the process of European economic integration has enjoyed success since the 1950s, cooperation in the field of space policy has lagged behind and was late to develop. This is striking, given that it "fitted perfectly with Jean Monnet's definition of an ideal area for the advancement of European integration" (Hörber, 2009b, p. 405), being too big for individual nation States but also offering a new field of politics comparable with nuclear research under EURATOM (Ibid.; Hörber, 2016a). In fact, in Europe space activities were originally conducted by individual States (Aliberti & Krasner, 2016), until in 1959 the model of the European Organisation for Nuclear Research (CERN) gave an impetus for the establishment of a European cooperative effort in space policy. This lead to the foundation of the European Launcher Development Organisation (ELDO) and the European Space Research Organisation (ESRO), until finally in 1975, ESA was created. Since then, ESA has been responsible for coordinating European space activities. The Single European Act (SEA) of 1986 is usually taken as a starting point marking the EC's/EU's development as a space actor, since it added research and development to the Community's competences (Reuter,

N.-L. Remuss, *Theorising Institutional Change: The Impact of the European Integration Process on the Development of Space Activities in Europe*, Springer Theses, https://doi.org/10.1007/978-3-319-95978-8_1

2007, p. 12; Marchisio, 2006). However, the Commission's interest in space policy goes back as far as 1970, when it participated in the European Space Conference (ESC) (Hobe, Kunzmann, & Reuter, 2006, p. 64; Madders & Thiebaut, 1992). This was the dominant forum for Europe's space community from 1967 to 1975 (Madders, 1997, p. 149) with the objective to harmonise space policies in Europe (Sheehan, 2007, p. 78). Since then, the EU has gradually increased its interest in space activities culminating in the inclusion of space as a shared competence between the EU and the Member States through Article 189 in the Lisbon Treaty of 2009. This created a legal basis for the EU's action in the space field. Since 2003, the EU and ESA worked together on the basis of the FA, which introduced the "Space Council", comprising both the EU Council of Ministers in form of the Competitiveness Council and the ESA Council, as a common decision-making body.

The introduction of the EU's flagship programmes Galileo, the European navigation programme, and Copernicus, the Global Monitoring for Environment and Security programme, underline the EU's commitment to the space area. Today's European space landscape is thus divided into three distinct levels:

(1) the EU;
(2) the intergovernmental organisations ESA and EUMETSAT; and
(3) the national space agencies.

With EU and ESA both acting in the field of European space activities with differing[1] but substantially "dual memberships" and partially overlapping competencies, questions regarding coherence, closer cooperation and the adaptability particularly with regard to the institutional arrangement gave rise to an academic and political debate about the future institutional set-up and division of labour between the EU and ESA. In this context four scenarios for an institutional set-up or a division of labour between EU and ESA have been debated:

(1) the cooperation model; i.e. improved cooperation between the two institutions
(2) the implementing versus political actor model; i.e. differentiation between ESA as the implementing and EU as the political actor
(3) the integration model; i.e. integration of the EU Commission in ESA's institutional structure and
(4) the accession model, i.e. the accession of ESA as an agency in the EU's institutional framework (cf. Rüttgers, 1989; Madders & Thiebaut, 1992; Reuter, 2007; Hobe et al., 2006; Kries, 2003; Hobe, Heinrich, Kerner, & Froehlich, 2009; Hörber, 2009a).

The European Commission itself has tasked the independent consultancy Robert Berger in 2013 to analyse the evolution of the EU-ESA relations, providing a cost-benefit assessment, with the objective to propose several future options. Similar to the options already proposed by several experts, Berger summarised the following four options:

---

[1]ESA embraces non-EU States such as Switzerland and Norway and extra-European States such as Canada.

(1) Option 1: Baseline scenario—No action is taken.
(2) Option 2: Improved cooperation under the "status quo"—ESA remains an independent intergovernmental organisation and there is a revision of the existing 2004 EU/ESA FA.
(3) Option 3: ESA remains an independent intergovernmental organisation within which new structures are created to deal with EU programmes in accordance with the EU financial regulation and laws; these structures within ESA would be accountable to the EU; an EU-Chamber is foreseen staffed by the EU Commission.
(4) Option 4: Transforming ESA into a EU agency while preserving ESA's successful intergovernmental features (European Commission, 2012, p. 4).

The EU's ambitions to incorporate space policy into community policies raise questions concerning the role of ESA being no longer the only European space actor. With the study of institutions lying at the heart of political sciences, the inter-institutional relationship between the EU and ESA seems to be a particularly interesting case.

While political science is often considered to be the study of institutions (cf. Steinmo, Thelen, & Longstreth, 1992, p. 3), the study of institutions is rather diverse, touching upon several academic disciplines, such as International Relations, International Law, International Political Economy, International History and European Studies (Rittberger & Zangl, 2006). Therein, the study of the EU seems to be a special case, proceeding in waves, with initial approaches such as Neofunctionalism and Intergovernmentalism focusing on explaining the path of integration (Koremenos, Lipson, & Snidal, 2001, p. 766) and more recent approaches analysing policy-making with (supranational) institutions and examining the process of institutional change (Caporaso, 1998, p. 2; Meunier & McNamara, 2007, p. 3). Since the mid-1980s institutions became central to the analysis when New Institutionalism was born, a diverse research programme, which can be summarised into three sub-approaches: Rational Choice Institutionalism, Historical institutionalism and Sociological Institutionalism.[2] Arguably, it is "more a perspective on politics than a fully developed theory" (Peterson & Shackleton, 2006, p. 5) and there is considerable confusion on how it differs from other approaches (Hall & Taylor, 1996b, p. 5). Its basic lesson is "institutions matter" (Rosamond, 2000, p. 113) and they are more than merely black boxes. Against this background it is more than striking that the definitions on "institutions" still vary (Rhodes, Binder, & Rockman, 2006, p. 7; Mearsheimer, 1994, p. 7) from "organisations", "social norms" (Héritier, 2007, p. 5) and "shared beliefs" to broader understandings such as "the foundation of social life" (Campbell, 2004, p. 1). The term is not only broadly defined (cf. Mearsheimer, 1994, p. 7) but also often used interchangeably with terms such as "organisations" and "regimes". Various schools use different definitions and even within one school different definitions are used (Boin, 2008, p. 89). Given that scholars do not agree on the definition of institutions, they also do not agree on how best to explain or understand the source of or the

[2]The best overview of the defining features, as well as the characteristic strengths and weaknesses of each can be found in Hall and Taylor (1996a, pp. 936–57) and Rhodes et al. (2006, p. xiii).

effects of international institutions and institutional change (Goldstein & Steinberg, 2010, p. xxiii). The European integration process and the institutions of the EU pose additional challenges on this already diverse existing theoretical framework.

1.1 Preview of the Argument

Among the disciplines mentioned above the question about the impact of the European integration process on existing institutional actors seems to be a case, which has so not been covered substantially so far. While, several authors tried to conceptualise the relationship between the EU and other intergovernmental organisations (in particular NATO, Council of Europe, OSCE) no structured debate can be identified.

At the same time the literature on institutional change seems to lack a structured approach, regarding the qualification and analysis of institutional change. Additionally, a theoretic framework making different cases of institutional change comparable, seems to be missing. Instead the existing literature on institutions has dealt with more general questions such as why States rely on international institutions as vehicles of cooperation, what attributes account for their use and how these characteristics set formal institutions apart from regimes and organisations. Additionally, scholars have recently begun considering concepts from the New Economics of Organisation (NEO), problems of delegation (cf. Bendor, Glazer, & Hammond, 2000), principal-agent aspects (cf. Edelmann, Milde, & Weimerskirch, 1998) as well as questions such as whether member governments are dominant and to what degree they exercise control over supranational agents (Doleys, 2000, p. 534).

This dissertation thus seeks to advance the debate on institutional change by proposing a more structure approach to the analysis of institutional change. In order to do so, existing theories are being analysed in terms of what they offer to the analysis of institutional change. Thereby, the evolving institutional arrangement between EU and ESA is considered a case study. Special attention is paid to the impact of the European integration process on existing institutional actors. Most scholars agree that no single theory can account adequately for everything (Egeberg, 2004, p. 199), as they offer only "a selective grasp on reality" (Héritier, 2007, p. 39) but are valuable "because they structure all observations" (Rosamond, 2000, p. 4). Thus, different theories are understood as complementary rather than competing, answering complementary aspects of the same puzzle. Such an approach is in line with the second movement in institutional analysis of the 1990s (Campbell, 2004, p. 4).

Following Keohane and Nye (2001, p. 50) it is analysed, which factors are most important in describing institutional change. It is argued that only the combination of these aspects forms a complete picture of institutional change. A process perspective is adopted, looking at the factors driving institutional change, the underlying causal processes and their outcomes, trying to theorise them (Héritier, 2007, pp. 7–8). The resulting aspects are presented as "facets of institutional change", which are proposed to structure the analysis of institutional change. They are applied to the observations of the development of space activities in Europe as a first test. This is in line with

Richardson's observation that the complexity of the EU policy process requires multiple models and the utilisation of "concepts from a range of models in order to at least describe the policy process accurately" (Richardson, 1996, p. 48).

1.1.1 Research Questions

Against this background, the "hot potato" currently under debate among experts in the space community is: "What will happen to ESA?" Can the provision of space policy only be conducted by one institution in Europe, i.e. is it a "zero-sum game" where (1) either the stronger institution grows and the more tasks the EU assumes, the more the role of ESA will diminish or (2) where the mere existence of ESA hampers the existence of an effective European space policy due to tensions arising out of duplication of structures, objectives and capabilities or (3) is the provision of space policy rather a win-win situation, where the EU and ESA will work together building up a coherent institutional set-up providing for a coherent European space policy avoiding unnecessary duplications? Since academics are no prognosticators, this dissertation cannot anticipate future developments. Alternatively, this dissertation conducts a historic analysis of the development of the inter-institutional relationship between the EU and ESA, guided by the question: **To what extent has ESA changed institutionally so far?**

At the same time the case of the inter-institutional relationship between the EU and ESA gives interesting insights into the following question: "**What is the impact of the European integration process on existing institutional actors?**". Additionally, in line with Doleys (2000), the question arises in **what situations governments can be expected to dominate policy-making**, and under **what conditions supranational actors can be expected to affect change independently** of and even in the face of opposition from governments (Doleys, 2000, p. 546). A starting point to the analysis of when, where, and how supranational institutions are able to implement their own agenda has been provided by Tallberg (2000, 2003). Accordingly, the initial delegated functions both condition the control mechanisms the principal may employ and shape the strategies agents may utilise, when pursuing their agenda (Tallberg, 2000, p. 846). It is thus focussed on the EU's Commission, when considering the case study.

According to Sandholtz & Stone Sweet (1998) institutions are in constant evolution, rarely emerge at once or "change as a single unit" (p. 16). What we perceive as institutional change can be the "normal" process of an institutional evolution. In line with this the question arises: "**Which stages of institutional development can be identified?**".

Considering the existing literature on institutional change, **which facets of institutional change can be identified with the objective to making different cases comparable? How can different cases of institutional change best be compared?** In line with Shepsle (1989), the selection, survival, adaptation and evolution of institutional practices are analysed (Shepsle, 1989, p. 133). Héritier (2007) has posed a similar question by asking: Which are the major rational institutionalist theories accounting for institutional change? What are the underlying arguments accounting for change and under which particular conditions are they particularly apt to explain a particular type of institutional change? (Héritier, 2007, p. 9). Following the objective to advance the debate on institutional change by proposing a more structured approach to the analysis of institutional change, I opt for a set of different questions, complementary to the ones identified before:

- What are the sources for institutional change?
- Which factors are most important in describing institutional change?
- Which outcomes of institutional change are possible?

The answers to these questions will provide building stones towards a structured approach to the analysis of institutional change, making different cases of institutional change comparable.

In this analysis institutions are thus understood to be both, the independent and dependent variables as they influence actors in their choices on the one hand, and are shaped by actors on the other hand.

1.1.2 Objectives

First, given the recurring academic and political debate concerning the future of ESA, an analysis of the historic development of space policy is needed. While several authors have already looked at this topic, there is still a need for a concise up-to-date analysis of the development of space collaboration in Europe from a political science perspective. Besides this, space as a policy domain has not been studied extensively yet.

Secondly, while theories of institutional change analyse catalysts for change of institutions, they do not seem to cover the impact of one institution on another one, but focus on States' influence instead. While the impact on existing institutional actors such as NATO has been the central to some academic discussion (cf. Kagan, 2004), there is still a need for an all-encompassing theoretic framework for the impact of the European integration process on existing institutional actors. This study seeks to provide a building block of such a theory by considering ESA as a case study, analysing when the Commission is able to advance its own agenda. Comparable cases would be the WEU (Bailes, 2011) and the Joint Aviation Authorities (JAA). In order to drive forward the debate about when supranational actors are able to advance their own agenda, this dissertation focuses on the role of the EU's Commission with its right of initiative, when analysing the case of ESA. Bearing in mind that the Commission is not a monolithic actor but composed of different Directorate-Generals (DGs), dealing with space policy, this dissertation analyses the Commission's documents only. An analysis of the roles of different DGs would extend beyond the realm of this research.[3]

Third, to make a theoretical contribution to our understanding of when supranational actors are able to advance their own agenda, it will be observed when the Commission is able to advance its own agenda. The development of space activities in Europe will serve as a case study for this.

Fourth, the study of institutions and institutional change seems to lack a structured approach. There is thus a need to develop a theory on institutional change. In line with Keohane and Nye (2001, p. 50) the objective of this dissertation is to identify factors, which are most important in describing institutional change. By considering the existing theoretic approaches to institutional change as complementary lenses rather than competing schools, "stages of institutional development" and "facets of institutional change", a facet being defined as "one of several parts" (Longman Dictionary of Contemporary English), being synonymous to "aspects", will be identified.

Against this background, the objectives of this study are fourfold.

- to provide an updated overview of the development of space policy in Europe
- to advance the debate about the impact of the European integration process on existing institutional actors
- to make a theoretical contribution to our understanding of when supranational actors are able to advance their own agenda
- to make a theoretical contribution to our understanding of institutional change.

1.1.3 Methodology and Approach

To be able to make a theoretical contribution to our understanding of institutional change, to advance the debate about the impact of the European integration process on existing institutional actors and to make a theoretical contribution to our understanding of when supranational actors are able to advance their own agenda, this dissertation will first review the state of the art of literature, dealing with institutional change and the European integration process. The second chapter will give an overview of the applicable tenants of European Integration Theories (i.e. Neofunctionalism, Intergovernmentalism and Liberal Intergovernmentalism, Supranational Governance), Theories of International Relations and New Institutionalism and analyse them according to factors perceived as most important in describing institutional change as called for by Keohane and Nye (2001, p. 50).

First, it is relied on Levy, Young and Zürn (1995) and Florensa (2004) to distinguish different stages of institutional developments, thereby claiming that institutions are in constant evolution (Sandholtz & Stone Sweet 1998, p. 16). Combining their suggestions, the process of institutional evolution seems to consist of five stages:

(1) agenda formation,

---

[3]For more information on the role of the Council of Ministers cf. Christiansen (2002).

(2) institutional choice (i.e. selection and design),
(3) operationalisation or evolution,
(4) redefinition, and
(5) crisis (Florensa, 2004, pp. 9–10; Levy et al., 1995, p. 282).

Additionally, there are two aspects to the evolution of an institution: first, the conditions under which institutions come about (agenda formation) and second, the process of competitive selection by relevant actors between different alternating institutional forms (institutional selection, choice or design) (cf. Hall & Taylor, 1996a, p. 945). It is suggested to analyse the process of institutional foundation in the light of several conditions, i.e. the "problem condition", the "cognitive condition", the "hegemonic condition", "windows of opportunity", and breaking points, which have been identified on the basis of Rittberger and Zangl (2006, p. 25), DiMaggio (1988, p. 14) and Olsen (1997, pp. 209–10, 217). Also, in line with Pollack (1996, p. 343), and Wendt (2001, pp. 1023–4) and Wallace (2000, pp. 43–4) it is suggested that the selection of a particular institution, is based on rational expectations about costs and benefits (Pollack, 1996, p. 434; Wendt, 2001, p. 1043) and normative considerations of what is appropriate (Wendt, 2001, pp. 1023–4). It will be interesting to see whether these can be identified in the case study or whether those institutions were chosen unintentionally as suggested by Heritier (2007, p. 1). Furthermore, it will be analysed in how far the founders of the institutions were constrained by the availability of institutional alternatives (Jupille, 2004, p. 1) and borrowed from prior institutions as suggested by Jupille (2004) and Scott (2008, p. 94). Additionally, it will be discussed in how far contextual factors, derived from the broad circumstances and functional factors, derived from some of the core functions of politics and demands to be met, as well as institutional factors [i.e. opportunities and constraints which influence in how far institutions are able to channel and to structure the behaviour of political actors from the participating countries (Wallace, 2000, pp. 43–4)] were relevant in the institutional selection of ESRO, ELDO and ESA.

Secondly, seeking to advance the debate on how to determine institutional change empirically, this dissertation proposes a set of facets of institutional change based on the above-mentioned existing literature. A facet is defined as "one of several parts" (Longman Dictionary of Contemporary English) being synonymous to "aspects". The following five facets of institutional change are proposed:

- the degree of change (incremental or revolutionary change; in which dimension it took place),
- the pace of institutional change (fast and slow changes),
- sources of institutional change (endogenous, exogenous),
- change agents, and
- forms of institutional change.

In combination with the previously mentioned phases of institutional evolution, the facets of institutional change are used to structure the observations of the case under discussion and to offer a structured approach to the analysis of institutional change in general.

Applying these to the case of ESA, the development of space activities in Europe is considered a case study of institutional change. In line with Jenkins-Smith and Sabatier (1993, pp. 1–9), Dudley and Richardson (1999, p. 244) and Hörber (2016b, p. 254) long-term historical policy studies are perceived as the only way to understand the long-lasting effects on policy systems such as institutional change. Thus, the development of European space activities is considered from the beginning of European space collaboration in 1959 until 2014. A chronological approach has been chosen as to identify changes. Bearing in mind the importance of the time frame for the analysis of institutional change (cf. Aminzade, 1992; Campbell, 2004, pp. 44–6), six phases of institutional development of space policy in Europe have been identified which, in line with Aminzade (1992) and Campbell (2004, pp. 44–6), were distinguished by critical events[4]:

1. Phase 1: The historic evolution of ESA (1959–1974).
2. Phase 2: From the founding of ESA to the entering of the EC in the space domain (1975–1986).
3. Phase 3: From the first steps of the EC to the ESA-EU Framework Agreement (1986–2004).
4. Phase 4: From the Framework Agreement to the Lisbon Treaty (2004–2009).
5. Phase 5: Post-Lisbon Treaty developments (2009–2012).
6. Phase 6: Post-ESA-EU Declaration of November 2012 (Post-2012).

Phase 1 and 2 can be separated by the foundation of ESA, which can be considered a critical event. Phase 2 and 3 are separated by the EU's first steps to enter the space arena through the introduction of research and development policy as European Communities' (EC) competences by the SEA of 1986. Many observers (Madders & Thiebaut, 1992; Reuter, 2007, p. 12) of the development of the EC and today's EU as a space actor mark this as the starting point. This development led to the formulation of the FA between the EU and ESA in October 2003, providing a "common basis and appropriate operational arrangements for efficient and mutually beneficial cooperation between ESA and the European Community" (Battrick, 2004, p. 5). It can be considered as another critical event separating phase 3 and 4. The FA resulted in the introduction of the common institution "Space Council" which characterises the fourth phase. The continuous rapprochement of the European Commission resulted in space being included as a shared competence in the Lisbon Treaty, signed in 2007 and entering force in 2009. This critical event marks the separation between the fourth and the fifth phase. Following the Lisbon Treaty, ESA and the EU were increasingly struggling to define their roles in the space policy field. ESA Member States thus considered the topic as part of their ESA Ministerial Conference in 2012 and issued a declaration entitled "Political Declaration towards the European Space Agency that best serves Europe" (ESA, 2012). This can be considered another critical mark, as ESA for the first time dealt publicly with the issue. It separates the fifth from the sixth phase.

[4] Accordingly, critical events may mark the appropriate time frames for analysis (e.g. electoral, budget, legislative or other cycles).

Combining the findings from the analysis of the state of the art of theories, explaining both the European integration process as well as institutional change, with the empirical findings, the six phases of the development of space policy in Europe are analysed in terms of the stages of institutional development and the five facets of institutional change.

Thus, the study builds on two blocks:

- the history of space policy in Europe.
- existing theoretical approaches to institutional change and the European integration process.

Both blocks are based on publicly available sources and desktop research only.

Thus, this dissertation is theory proposing, as it seeks to make a theoretical contribution to our understanding of institutional change, to advance the debate about the impact of the European integration process on existing institutional actors and to make a theoretical contribution to our understanding of when supranational actors are able to advance their own agenda. At the same time, it is historical explanatory, using the proposed institutional facets to explain institutional change considering the example of ESA, including a large empirical section. It can also be considered as being theory-testing as one of the conclusions will be, in how far further differentiation has to be conducted regarding the facets of institutional change as identified. Some of which might be too broad to distinguish different cases of institutional change.

1.1.4 Conclusions

By considering the existing theoretic approaches to institutional changes as complementary lenses rather than competing schools, this dissertation proposes to structure the analysis of institutional change, by relying on a combination of stages of institutional development and facets of institutional change. This answers the research question "When or why do institutions change?" and "which forms institutional change can take". Applying such an approach to different cases of institutional change does not only shed light on the case under discussion but also makes different cases comparable. Considering the development of ESA as a case study for institutional change, this dissertation highlights the advantages and disadvantages of such an approach.

Additionally, several claims of the existing literature will serve as areas of particular theoretical interest. Those are:

- **Precedence of States acting as change agents in scientific areas**: In policy areas which matter to national sovereignty, therefore being a symbol of scientific and technical capability and component of military strategy, States act as change agents presiding over other change agents. Thus, it is expected that States are the main change agents activating institutional change in all six phases under discussion, since space policy is a matter of national sovereignty, being a symbol of scientific and technical capability, an index of political power and a component of military strategy, and thus governed by States.

- **Precedence of individuals acting as change agents in times of uncertainty**: In contrast to the claim of many theoretic schools (cf. Intergovernmentalism, Realism, Neorealism, Neoliberal institutionalism etc.) about the process of institutional foundation being solely characterised by intergovernmental bargaining, it is argued that in times of uncertainty or when highly technical issues are being discussed skilful individuals with high social capital and the ability to access decision makers are able to advance their own agenda. This way, they act as change agents or can take the lead in founding institutions. Since space policy is of highly technical nature, it is expected that individuals could act as change agents in all phases under discussion (Haas, 1992, p. 3; Richardson, 1996, p. 40).
- **Precedence of groups acting as change agents in scientific areas and times of uncertainty**: Also, in times of uncertainty or when highly technical issues are being discussed policy-makers' demand information, which can be provided by expert groups such as epistemic communities (Haas, 1992, p. 12). Thus, in times of uncertainty, groups cause institutions to change. It is perceived as likely, in line with Haas (1992, p. 12) that this study finds proofs for groups acting as change agents, as space policy is a policy area, which is highly technical.
- **Precedence of supranational institutions acting as change agents in areas, where they have great authority**: According to theoretic accounts such as Neo-functionalism, Supranational Governance (cf. Hix, Pierson), Rational Choice Institutionalism supranational institutions exert significant independent influence on institutional and policy outcomes (Hix, 2005, p. 16), act as driving forces (Pierson, 1998, p. 29) and play an autonomous role (Andersen, Eliassen, & Sitter, 2001, p. 28) and can act as change agents. Supranational institutions such as the Commission advance its own agenda regarding topics where they have a great authority, legitimacy, or "institutional standing" and "institutional embeddedness" (Conner, 2010, p. ix). Also, where they have the control of information and knowledge, when they are able to abuse information asymmetries, to frame or shape a proposal in such a way that they are securing approval even in the light of Member State(s)'s opposition, or when they act together with powerful allies such as States. Given that both ESA and the EU operate in the same geographic realm, it is expected that institutions of the EU, i.e. the Commission as policy entrepreneur, will be found to have acted as a change agent.

In line with these areas of interest, the following conclusions are expected:

- **Pace of institutional change**: The pace of institutional change is distinguished in slow, incremental institutional change through small adjustments, gradual or evolutionary in a continuous manner and rapid revolutionary institutional change through large steps in a discontinuous manner, resulting in a "steep transformation", breakdown and replacement of old institutional arrangements (Campbell, 2004, p. 33; Madariaga, 2011, p. 17, 29; Roland, 2004, pp. 116–7). It is expected in line with Historical Institutionalists (i.e. North, 1990, p. 73; March & Olsen,

1989, p. 169; Pollack, 1996, p. 441; Campbell, 2004, p. 48) that ESA has mainly undergone incremental change.

- **Sources of institutional change**: Distinguishing between endogenous sources of institutional change (i.e. Greif & Laitin, 1994; Streeck & Thelen, 2005; Young, 2010) and exogenous (i.e. Pollack, 1996, p. 44; Krasner, 1988, p. 67; Scott, 2008; Steinmo et al., 1992; Suzuki, 2003). Endogenous sources of institutional change are self-undermining processes (Greif & Laitin, 1994, p. 634), institutional design (Héritier, 2007; Streeck & Thelen, 2005, p. 19) or internal contradictions (Young, 2010, p. 174), ideological innovation, learning processes through the acquisition of knowledge as well as skills and changes in tastes or preferences of the actors involved (Florensa, 2004, p. 6). Exogenous sources of institutional change, are "critical junctures or punctuation" (cf. Historical Institutionalism), i.e. changes in the attribution of natural resources, extreme climatic events or natural disasters,[5] changing preferences of supportive/influencing governments (Florensa, 2004, p. 18; Pollack, 1996, p. 441), which might result from unfulfilled expectations, shifts in power distributions of previously supportive coalitions or the appearance of new actors (Pollack, 1996, p. 438; Scott, 2008, p. 198), changing voter preferences, leading to new political alignments, lobby of business interest (Scott, 2008, p. 196), broad changes in the policy environment of social, political or economic nature (Steinmo et al., 1992, p. 16; Florensa, 2004, p. 18; Pollack, 1996, p. 438)[6] and improvements in the quality of information about either the operation of the institution or the environment within which it operates (Pollack, 1996, p. 438). Both endogenous and exogenous sources are expected to be found in all phases under analysis, with the European integration process being the main exogenous sources. It is also hoped to find answers to the question under what conditions endogenous or exogenous sources preside.
- **Change agents**: Suggesting that sources of institutional change "have to be activated by agents willing to generate change" (Madariaga, 2011, p. 24), such as States (and their ministries, representatives), individuals, groups (interest groups and communities of experts) or institutions (administrative staff of international organisations; parliamentary assemblies of international organisation) (Rittberger & Zangl, 2006, p. 78), it is expected to find different change agents and not solely States as sources for institutional change for the case under discussion.

The dissertation also seeks to advance the debate about the impact of the European integration process on existing institutional actors, thereby making a theoretical contribution to our understanding of when supranational actors are able to advance their own agenda.

---

[5]For more information on "critical junctures" refer to the following literature cf. Florensa (2004, p. 18), Christiansen and Vanhoonacker (2008), Campbell (2004, p. 26), Hall and Taylor (1996a, p. 942), Krasner (1984, p. 235) in Ikenberry (1994, p. 9), Horak (2007, p. 21), March and Olsen (2006, p. 12).

[6]For example changes in the markets, technological innovations or legislative changes.

1.2 Relevance

Until recently space policy and ESA "attracted very little serious and thoughtful attention" (Hörber, 2009b, p. 412) from politicians or the general public, being regarded as a field of experts and Sci-Fi geeks. "This has changed in the past decade (…) "with the potential benefits of space exploration" (Ibid.) being "discovered by politicians and economists alike" (Ibid.).

Space policy is a particularly interesting case. First, ESA was founded as an independent institution entirely separate from the European integration process that resulted in the EU. Thus, as Hörber (2016c) put it, "the fact that there is now close cooperation with the EU is significant and worthy of investigation" (Hörber, 2016a, p. 54). Secondly, space policy has the potential to foster what European integration has promised: peace and prosperity, while it will always have the potential to attract "the mind and hearts of many" (Hörber, 2016, p. xxii). Directly or indirectly space is relevant to every EU policy area (Hörber, 2016c, p. 11). Space is very much at the cutting edge of current EU policy developments. Recent phenomena such as climate change, rapid developments in crisis management and prevention etc. all need space capabilities to "manage" them (Ibid.). As Hörber (2016b) mentioned, space policy itself can also be considered a driving force of European integration. As such it has not yet been sufficiently understood because "space has not been studied as a policy domain as have defence, environment, transport or climate change (p. 254)". The strategic potential of space for the EU is likely to grow, given the many possible future fields of engagements such as commercial launches, missions to Mars, Moon research or a Moon village. The industrial and economic potential of this area of activity is further motivation to strengthen the common frame between the ESA and the EU (Hörber, 2016a, p. 61). Moreover, space policy has the potential to foster a European identity (cf. Rathgeber & Remuss, 2009; Remuss, 2011).

At the same time the question on how to explain both institutional stability and change is a central concern in political science today (Greif & Laitin, 1994, p. 634). While we know a lot about the creation of institutions, relatively little research has been conducted about their transformation once they have been put into place (Héritier, 2007, p. 1), as has been explained by Héritier. This is particularly striking, as institutional change can be seen as a quasi-permanent process in the EU. The causes and outcomes of institutional change are particularly interesting, as they raise a number of conceptual, theoretical and empirical questions (Héritier, 2007, p. 1). In line with this Shepsle (1989) underlined the need to understand the selection, survival, adaptation and evolution of institutional practices (Shepsle, 1989, p. 133). As Campbell (2004, p. 31) suggested, if we want to specify conditions under which different types of changes are more likely to occur, we need to be able to identify different types of change first. Sigalas (2016), who analysed the role of the European Parliament in framing space policy, recently underlined the need for "a fully-fledged theory of change" (p. 77), when he tried to explain exactly why the EP's justification frames changed (Ibid.). This proofs the relevance of this study.

Additionally, the evolution of the EU as an international actor, integrating ever more policy fields, led to a process of institutional change of already existing international organisations. Some were included as EU agencies into the community framework (i.e. the Western European Union (WEU) Satellite Centre (cf. Davara, 2000), the WEU Security Institute (cf. Council of the European Union, 2006), the Western European Armaments Organisation (WEAO) and the Western European Armaments Group (WEAG) (Bailes, 2011). Others were dissolved (the WEU, the JAA), and a third group of institutional actors managed to co-exist (CERN, CEPT). Against this background the question arises, how to account for the different institutional outcomes resulting from the impact of the European integration process. This phenomenon is particularly interesting from the political sciences and the institutional change perspective. The development of space policy in Europe serves as a case study for such a research on the impact of the European integration process on other international organisations, thereby advancing the general field of study on the process of institutional change.

At the same time Pollack (1997a) concludes "the precise causal role of the European Commission in the process of European policy-making and European integration remains theoretically contested and empirically unmapped" (Pollack, 1997b, p. 111). While the Commission's precise role has remained in dispute, the very notion of it acting as an agent has become "a semantic bone of contention" (Pollack, 1997b, p. 111) between Neofunctionalism and Intergovernmentalism. In line with this, this dissertation analyses the role of the Commission as a change agent.

Additionally, in line with Doleys (2000, p. 546) the question arises in what situations governments can be expected to dominate policy-making, and under what conditions supranational actors can be expected to affect change independently of and even in the face of opposition from governments.

1.3 State of the Art and Innovation

Several authors have perceived ESA and the inter-institutional relationship or space governance structure in Europe worth the analysis.

First, Hutter (1991) looked at the factors characterising European space policy. He concluded that apart from national and European considerations, characteristics of the policy field itself, i.e. its changes as well as the technological and scientific preconditions for space applications, influence European space policy (Hutter, 1991, p. 5). Additionally, the international context, i.e. the explicit and implicit influence of the U.S. and Russia influence(d) the content of the European space policy (Ibid.).

Secondly, the law faculty of the University of Cologne (cf. Hobe, Schmidt-Tedd, & Schrogl, 2006; Hobe et al., 2006) has conducted a project on the legal framework conditions for a future coherent structure of the European space policy. Such a project was deemed relevant, given the proceeding of the European integration process at the beginning of the twenty-first century. While highlighting four different scenarios for the future of the institutional structure of European space policy, the authors weight several models against each other, but remain undecided regarding the future of ESA. Most authors agree on the necessity to review the existing set-up, but limit themselves to legal questions and neither touch upon political nor theoretical questions. Among these Reuter (2007) analysed, whether it would be legally possible to make ESA the EU's space agency (Reuter, 2007). Moreover, Aliberti and Krasner (2016) recently compared the existing rules and regulations of international space activities with the ones of Antarctic and the oceans, again taking a legal perspective.

Third, Kevin Madders (1997) in his book "New Force as a New Frontier" aimed "to present the European space field organically, especially through concentrating on its main actors and what they have done, as well as on the issues they and the sector as a whole face today" (Madders, 1997, p. 587). While providing a good overview of the historic development of space collaboration in Europe, Madders did not analyse the development from the perspective of institutional change or European integration theories. His work provides a historic description, lacking theoretical analysis.

Fourth, the first theoretic analysis of European space collaboration was provided in 2003 by Suzuki. He aimed to provide an analytical framework for understanding the dynamics of international collaboration and to place European space collaboration in a historical and analytical context (Suzuki, 2003, p. 1). He sought to explain changes of decisions, negotiations and institutional frameworks with his concept of policy logics (Suzuki, 2003, p. iix). Accordingly, European space collaboration is a consequence of a dynamic process of interactions between the subjective understanding of policy-makers about space activities and the external environment in which the U.S. has been the dominant actor (Suzuki, 2003, p. 1). Suzuki (2003) criticised the neo-realist framework with its three assumptions (rational State, protection of autonomy, and centralisation of intergovernmental bargains) as insufficient for the analysis of European space collaboration and instead relied on historical and rational choice institutionalism. He challenged the "unitary State" assumption of Neo-Realism and introduced the two-level games approach, separating the decision-making process into international negotiations and the domestic ratification process. Accordingly, a plurality of both actors and objectives within national governments exists, which imposes constraints upon domestic actors in international negotiations. Suzuki consequently introduced his concept of "policy-logic-analysis", which focuses on the logics that actors may follow for defining their long-term interest and goals, their policy conduct and behaviour, as well as a set of values, which could provide criteria for judging policy performance. At the same time he focused on the role of institutions at different levels of action and on the implications for the behaviour of the actors (Suzuki, 2003, p. 7). He did not aim to create a new theory for the study of international relations or European integration but wanted to understand why and how European space collaboration emerged and evolved in the course of history, thereby identifying driving forces behind changes. To this end, he made a first attempt in providing an analytical and non-theoretical framework of the nature of European collaboration (Ibid., p. 8). Given that his work was published more than ten years ago, there is a need for a concise up-to-date analysis of the development of space collaboration in Europe. Also, his proposal of "policy logics" can be seen as only part of the puzzle missing many more facets of the analysis of institutional change, such as pace of change and different possible outcomes of change.

Fifth, most recently, the edited book of Hörber and Stephenson entitled "European Space Policy—European integration and the final frontier", which was published in 2016, applied Framing Theory to the case of space policy. Arguing that Framing Theory has become increasingly popular in recent European Studies, it aims to evaluate the state of European integration at the current time and the part that European Space Policy (ESP) could play in furthering this process. It considers ESP as a tool for the analysis of European integration and presents space policy as an original and timely lens for studying developments in European policy-making. The authors perceive space policy as an important contributor to—and catalyst for—the European integration process. By analysing the European integration process, the authors sought to develop ideas about future strategies for fostering that process through an ESP (Hörber, 2016d, p. 1). Relying on Framing Theory, the authors identified particular frames operating in the discourse of different actors. By examining actors and institutional interests, they provide the reader with an explanation of how ESP operates through a discursive lens (Hörber, 2016c, p. 4). It is argued that the European integration process needs a new direction and that ESP might have the potential to renew the idealistic forces of early post-war integration and make twenty-first century Europeans reach for new horizons (Hörber, 2016c, pp. 5–6). It contains an in-depth overview of framing theory, making its applicability to European space policy clear (Hörber, 2016c, p. 7). The authors analysed the institutions with Hörber examining the role of ESA, Sigalas that of the European Parliament, Köpping Athanasopoulos that of the Council, Marta & Stephenson that of the Commission and Forganni that of the European Court of Justice. Such a comprehensive analysis of the roles of the EU and other international institutions in defining space policy has not been published before. All authors used framing theory as an analytical tool in their chapters (Hörber, 2016b, p. 253). Köpping Anthanasopoulos (2016, p. 91) for example found three distinct rational frames for space policy which became recurrent in the Council's publications on space. These are the security frame, the independence frame and the prosperity frame. He concluded that space policy is an example of the interplay between the Commission as the frame-creator and the Council as the frame selector (p. 84). The focus on Framing Theory is a good starting point and in fact the first analysis of space as a European policy area. However frames were highlighted in terms of discursive analysis of existing policy papers of the actors under discussion, lacking conclusions regarding how changes in these frames came about. This is highlighted by Sigalas (2016) who analysed the role of the European Parliament in framing space policy, recently concluding that "without a fully-fledged theory of change, it is not possible to explain exactly why the EP's justification frames changed" (p. 77). This proofs the relevance of this study.

As Hörber (2009a) put it, the literature on the European space effort is quite limited. While reference can be found in books dealing mainly with security-related issues, a structured analysis of the European space effort is missing (Hörber, 2009b, p. 406). Thus, while taking the existing literature as a starting point, this study will go back to the theoretic accounts on institutional change and the historic sources of European space collaboration and to look at:

- different facets of institutional change making a theoretical contribution to our understanding of institutional change and becoming able to structure the analysis of institutional change and to compare different cases of institutional change
- the impact of the European integration process on the development of space activities in Europe, taking ESA as a case for the analysis of the impact of the European integration process on existing institutional actors.

As has been mentioned before, the study of institutions lies at the heart of political sciences and has been developed in recognisable waves. The state of the art of theoretic work on institutional change will be considered in a separate chapter. The literature is diverse, differs in definitions and does not propose a structured approach to the study of institutional change and the case of the European integration process as well as its impact on existing institutional actors. The only existing structured approach is offered by Heritier (2007). Focussing on Rational Choice Institutionalism, she considers changes in the European institutions i.e. European Parliament, Presidency of the Council and the Commission as empirical cases. Her work was taken as a starting point and extended by more facets of institutional change. She distinguishes between a process perspective, focussing on the factors driving institutional change, and a structural perspective, looking at the actors involved (Héritier, 2007, pp. 7–8). This dissertation argues that to gain a complete picture of institutional change both perspectives are necessary and have to be combined.

In addition to the existing literature on the EU-ESA inter-institutional relationship, several authors have already dealt with the phenomenon of institutional change Several important puzzles however remain.

The literature on the role of the Commission can be divided into two main branches, the first dealing with it as an international administration, the second considering it as a black box investigating its role among the other EU institutions (Kassim et al. 2016, p. 17). Much of the agenda-setting literature for example pays less attention to the role of the decision-making structure in explaining how different interests are aggregated, while the bureaucratic politics school of thought emphasises the administrative structure of the European Commission (Hartlapp, Metz, & Rauh, 2013, p. 425). A structured approach, regarding the conditions under which supranational actors can be expected to affect change independently of and even in the face of opposition from governments has so far not been offered (cf. Doleys, 2000, p. 546). A starting point to the analysis of when, where, and how supranational institutions can implement their own agenda has been provided by Tallberg (2000, 2003). By relying on Rational Choice Institutionalism and the principle-agent-approach he explains the variations in autonomy of supranational institutions by the initial functions delegated to them. These condition the control mechanisms of States and shape the strategies they may utilise to advance their own agenda (Tallberg, 2000, p. 846). He concluded that the while the Commission is equipped with the right of initiative in the legislative procedure, it must still mobilise Member States to implement its own agenda, as is it subject to an elaborate system of control mechanisms (Tallberg, 2000,

p. 861). Thus, independent influence of supranational institutions is conditioned by Member States' means of monitoring and sanctioning of the supranational institution (Tallberg, 2003, p. 128). As an incentive for supranational influence of the EU institutions, he observed the desire to secure State adherence to EU rules (Tallberg, 2003, p. 130). He also found the Commission's independent influence limited at moments of treaty revision (Tallberg, 2003, p. 131).

1.4 Outline

The first chapter takes stock of the existing theoretic accounts dealing with institutional change. Thereby the stages of institutional development as well as different facets of institutional change will be distinguished, to offer a more structured approach to institutional change. Following Keohane and Nye (2001, p. 50) it is analysed, which factors are most important in describing institutional change.

Chapter 2 recalls the history of the development of space activities in Europe. In order to do so, six phases of institutional development of space policy in Europe will be identified. While adopting a chronological approach to identify historical changes, the phases were separated according to institutional differences.

Chapter 3 seeks to combine the methodological and the empirical part of this dissertation. Therefore, the six phases of the development of space policy in Europe will be analysed in terms of stages of institutional development and facets of institutional change as identified. Particular attention is paid to the impact of the European integration process as a source for institutional change. Furthermore, the objective is to find out in how far the facets of institutional change are suited to structure the debate on institutional change. The case of the development of cooperation in space policy in Europe is used as a testing ground.

The final chapter will summarise the conclusions and offer an outlook, assessing the proposed approach in terms of generalisability.

References

Aliberti, M., & Krasner, S. D. (2016). Governance in space. In C. Al-Ekabi, B. Baranes, P. Hulsroj, & A. Lahcen (Eds.), *Yearbook on space policy 2014* (pp. 143–166). Vienna: Springer.

Aminzade, R. (1992). Historical sociology and time. *Sociological Methods & Research, 20,* 456–480.

Andersen, S. S., Eliassen, K. A., & Sitter, N. (2001). Formal processes: EU institutions and actors. In S. S. Andersen & K. A. Eliassen (Eds.), *Making policy in Europe* (pp. 20–43). London: SAGE Publisher.

Bailes, A. J. K. (2011). *Death of an institution. The end for western European union, a future for European defence? Egmont Paper 46*. Retrieved from http://aei.pitt.edu/32322/1/ep46.pdf. Accessed March 13, 2017.

Battrick, B. (Ed.). (2004). *The European space sector in a global context—ESA's annual analysis (BR-222)*. Retrieved from http://www.esa.int/esapub/br/br222/br222.pdf. Accessed March 19, 2017.

Bendor, J., Glazer, A., & Hammond, T. H. (2000). *Theories of delegation in political science*. Retrieved from https://www.gsb.stanford.edu/faculty-research/working-papers/theories-delegation-political-science. Accessed January 18, 2012.

Boin, A. (2008). Mapping trends in the study of political institutions. *International Studies Review, 10,* 87–92.

Campbell, J. (2004). *Institutional change and globalization*. Princeton and Oxford: Princeton University Press.

Caporaso, J. (1998). Regional integration theory: Understanding our past and anticipating our future. *Journal of European Public Policy, 5*(1), 1–16.

Christiansen, T. (2002). Out of the shadows: The general secretariat of the council of ministers. *The Journal of Legislative Studies, 8*(4), 80–97.

Christiansen, T., & Vanhoonacker, S. (2008). At a critical juncture? Change and continuity in the institutional development of the council secretariat. *West European Politics, 31*(4), 751–770.

Conner, A. (2010). Introduction: International institutions and effectiveness—Achieving cooperation. In J. L. Goldstein & R. H. Steinberg (Eds.), *International institutions—Vol. II consequences: When, where and why international institutions are effective* (pp. iiix–x). London: SAGE.

Council of the European Union. (2006). *Council joint action amending joint action 2001/554/CFSP on the establishment of a European union institute for security studies (2006/1002/CFSP)*. Brussels: European Communities.

Davara, F. (2000). The future of the WEU Torrejon Centre. *Air & Space Europe, 2*(4), 24–25.

DiMaggio, P. (1988). Interest and agency in institutional theory. In L. G. Zucker (Ed.), *Institutional patterns and organizations: Culture and environment* (pp. 3–23). Cambridge: Ballinger Pub.

Doleys, T. (2000). Member states and the European commission: Theoretical insights from the new economics of organization. *Journal of European Public Policy, 7*(4), 532–553.

Dudley, G., & Richardson, J. (1999). Competing advocacy coalitions and the process of "frame reflection": A longitudinal analysis of EU steel policy. *Journal of European Public Policy, 6*(2), 225–248.

Edelmann, R., Milde, H., & Weimerskirch, P. (1998). Agency-Beziehungen und Kontrakt-Design: Problem, Lösung, Beispiel. *Kredit und Kapital, 1,* 1–27.

Egeberg, M. (2004). An organisational approach to European integration: Outline of a complementary perspective. *European Journal of Political Research, 43,* 199–219.

ESA. (2012). *Political declaration towards the European space agency that best serves Europe* (ESA/C-M/CCXXXIV/Res.4 (final)). Naples: ESA.

European Commission. (2012). *Establishing appropriate relations between the EU and the European space agency (COM (2012) 671 final)*. Brussels: European Communities.

Florensa, M. C. (2004). *Institutional stability and change. A logic sequence for studying institutional dynamics*. Paper presented at the Tenth Biennial Conference of the International Association for the Study of Common Property (IASCP). Oaxaca, Mexico. Retrieved from https://dlc.dlib.indiana.edu/dlc/bitstream/handle/10535/1823/CostejaFlorensa_Institutional_040525_Paper220.pdf?sequence=1. Accessed March 13, 2017.

Gaubert, A. (2006). Is there really any duplication in Europe's space activities? *Space Policy, 22*(1), 1–2.

Goldstein, J. L. & Steinberg, R. H. (2010). Introduction. In J. L. Goldstein, & R. H. Steinberg (Eds.), *International Institutions, Vol. I: Causes* (pp. xvii–xxxi). London: SAGE.

Greif, A., & Laitin, D. D. (1994). A theory of endogenous institutional change. *American Political Science Review, 98*(4), 633–652.

Haas, P. M. (1992). Introduction: Epistemic communities and international policy coordination. *International Organization, 46*(1), 1–35.

Hall, P. A., & Taylor, R. C. R. (1996a). *Political science and the three new institutionalisms* (pp. 936–957). XLIV: Political Studies.

Hall, P. A., & Taylor, R. C. R. (1996b). *Political science and the three new institutionalism*. Retrieved from http://www.mpifg.de/pu/mpifg_dp/dp96-6.pdf. Accessed March 13, 2017.

Hartlapp, M., Metz, J., & Rauh, C. (2013). Linking agenda setting to coordination structures: Bureaucratic politics inside the European commission. *Journals of European Integration, 35*(4), 425–441.

Héritier, A. (2007). *Explaining institutional change in Europe*. New York: OUP.

Hix, S. (2005). *The political system of the European union*. Houndmills: Palgrave.

Hobe, S., Heinrich, O., Kerner, I., & Froehlich, A. (2009). *Entwicklung der Europäischen Weltraumagentur als „implementing agency" der Europäischen Union: Rechtsrahmen und Anpassungserfordernisse*. Kölner Schriften zum Internationalen und Europäischen Recht. Band 17. Berlin: LIT Verlag.

Hobe, S., Kunzmann, K., & Reuter, T. (2006). *Forschungsbericht ESA—EU: Rechtliche Rahmenbedingungen einer zukünftigen kohärenten Struktur der europäischen Raumfahrt*. Kölner Schriften zum Internationalen und Europäischen Recht Bd. 13. Berlin: LIT Verlag.

Horak, M. (2007). *Governing the post-communist city: Institutions and democratic development in Prague*. Toronto: University of Toronto.

Hörber, T. (2009a). The European space agency and the European union: The next step on the road to the stars. *Journal of Contemporary European Research, 5*(3), 405–414.

Hörber, T. (2009b). ESA + EU: Ideology or pragmatic task sharing? *Space Policy, 25*, 206–208.

Hörber, T. (2016a). The European space agency and the European union. In T. Hörber & P. Stephenson (Eds.), *European space policy* (pp. 53–65). London, New York: Routledge.

Hörber, T. (2016b). Conclusion: To boldly go where no one has gone before. In T. Hörber & P. Stephenson (Eds.), *European Space Policy* (pp. 253–263). London and New York: Routledge.

Hörber, T. (2016c). Foreword. In T. Hörber & P. Stephenson (Eds.), *European space policy* (pp. xx–xxiii). London and New York: Routledge.

Hörber, T. (2016d). Introduction—Towards a European space policy. In T. Hörber & P. Stephenson (Eds.), *European space policy* (pp. 1–12). London and New York: Routledge.

Hutter, T. (1991). *Charakteristika und Bestimmungsfaktoren der europäischen Raumfahrtpolitik*. Dissertation zur Erlangung des akademischen Grades Doktor der Sozialwissenschaft an der Fakultät für Sozial- und Verhaltenswissenschaften der Eberhard-Karls-Universität Tübingen.

Ikenberry, G. J. (1994). *History's heavy hand: Institutions and the politics of the state*. Paper prepared for a conference on "New Perspectives on Institutions", University of Maryland. Retrieved from: https://scholar.princeton.edu/sites/default/files/HistorysHeavyHand_0.pdf. Accessed March 13, 2017.

Jenkins-Smith, H. C., & Sabatier, P. A. (1993). The study of public policy processes. In H. C. Jenkins-Smith & P. A. Sabatier (Eds.), *Policy change and learning. An advocacy coalition approach* (pp. 1–9). Boulder: Westview Press.

Jupille, J. (2004). *Procedural politics: Issues, influence, and institutional choice in the European union*. Cambridge: Cambridge University Press.

Kagan, R. (2004). *Paradise and power. America and Europe in the new world order*. London: Atlantic Books.

Kassim, H., Connolly, S., Dehousse, D., Rozenberg, O., & Bendjaballah, S. (2016). Managing the house: The presidency, agenda control and policy activism in the European commission. *Journal of European Public Policy*, http://dx.doi.org/10.1080/13501763.2016.1154590, pp. 1–22.

Keohane, R., & Nye, J. (2001). *Power and interdependence*. New York: Longman.

Köpping Anthanasopoulos, H. (2016). Europe's wilderness—The council's frames on space policy. In T. Hörber & P. Stephenson (Eds.), *European space policy* (pp. 82–97). London and New York: Routledge.

Koremenos, B., Lipson, C., & Snidal, D. (2001). The rational design of international institutions. *International Organization, 55*(4) (Autumn 2001), 761–799.

Krasner, S. (1988). Sovereignty—An institutional perspective". *Comparative Political Studies, 21*(1), 66–94.

Levy, M. A., Young, O. R., & Zürn, M. (1995). The study of international regimes. *European Journal of International Relations, 1*, 267–330.

Madariaga, A. (2011). *Patterns of institutional change and external competitiveness in neoliberal and dependent political economies. The cases of Chile and Estonia*. Budapest: Central European University.

Madders, K. (1997). *A new force at a new frontier: Europe's development in the space field in the light of its main actors, policies, law and activities from its beginning up to the present*. Cambridge: Cambridge University Press.

Madders, K., & Thiebaut, W. M. (1992). Two Europe in one space: The evolution of relations between the European space agency and the European community in space affairs. *Journal of Space Law, 20*(2), 117–132.

March, J. G., & Olsen, J. P. (1989). *Rediscovering institutions: The organizational basis of politics*. New York: The Free Press.

March, J. G., & Olsen, J. P. (2006). Elaborating the 'New Institutionalism'. In R. A. W. Rhodes, S. A. Binder, & B. A. Rockman (Eds.), *The Oxford handbook of political institutions* (pp. 3–20). Oxford: OUP.

Marchisio, S. (2006). Proposals for an institutional realignment of the European space sector. In S. Hobe, B. Schmidt-Tedd, & K.-U. Schrogl (Eds.), *Project 2001 plus—global and European challenges for air and space law* (pp. 193–204). Köln: Heymanns.

Mearsheimer, J. (1994). The false promise of international institutions. *International Security, 19*(3), 5–49.

Meunier, S., & McNamara, K. R. (2007). *Making history: European integration and institutional change at fifty*. Oxford: OUP.

North, D. C. (1990). *Institutions, institutional change and economic performance*. Cambridge: Cambridge University Press.

Olsen, J. P. (1997). Institutional design in democratic context. *The Journal of Political Philosophy, 5*(3), 203–229.

Peterson, J., & Shackleton, M. (2006). The EU's Institutions: An overview. In J. Peterson & M. Shackleton (Eds.), *The institutions of the European union* (pp. 1–16). Oxford: OUP.

Pierson, P. (1998). The path to European integration: A historical institutionalist perspective. In W. Sandholtz & A. Stone Sweet (Eds.), *European integration and supranational governance* (pp. 27–58). Oxford: OUP.

Pollack, M. A. (1996). The new institutionalism and EC governance: The promise and limits of institutional analysis. *Governance: An International Journal of Policy, Administration, and Institutions, 9*(4), 429–458.

Pollack, M. A. (1997a). Delegation, agency, and agenda setting in the European community. *International Organization, 51*(1), 99–134.

Pollack, M. A. (1997b). The Commission as an agent. In N. Nugent (Ed.), *At the heart of the union: Studies of the European commission* (pp. 111–130). Houndmills: Macmillan.

Rathgeber, W., & Remuss, N. (2009). *Space security—A formative role and principled identity for Europe* (ESPI Report 16). Wien: ESPI.

Remuss, N. (2011). Meeting Europe's external and internal security through space applications—Security as an element for European identity. In B. Baranes & K.-U. Schrogl (Eds.), *European identity through space* (pp. 132–144). Vienna: Springer.

Reuter, T. (2007). *Die ESA als Raumfahrtagentur der Europäischen Union: Rechtliche Rahmenbedingungen für eine institutionelle Neuausrichtung der europäischen Raumfahrt*. Köln: Carl Heymanns Verlag.

Rhodes, R. A. W. (2006). Old institutionalism. In R. A. W. Rhodes, S. A. Binder, & B. A. Rockman (Eds.), *The Oxford handbook of political institutions* (pp. 90–108). Oxford: OUP.

Richardson, J. (1996). Actor-based models of national and EU policy making. In H. Kassim & A. Menon (Eds.), *The European union and national industrial policy* (pp. 26–51). London: Routledge.

Rittberger, V., & Zangl, B. (2006). *International organization: Polity, politics and policies*. New York: Palgrave Macmillan.

Roland, G. (2004). Understanding institutional change: Fast-moving and slow-moving institutions. *Studies in Comparative International Development, 38*(4), 109–131.

Rosamond, B. (2000). *Theories of European integration*. Houndmills: Palgrave.

Rüttgers, J. (1989). *Europas Wege in den Weltraum: Programme—Proteste—Prognosen*. Frankfurt am Main: Umschau Verlag.

Schrogl, K. (1993). Die Europäische Gemeinschaft als Magnet—Ihre Anziehungskraft auf die europäischen Fachorganisationen ESA und CEPT. *Zeitschrift für Internationale Politik, 18*, 525–532.

Scott, R. W. (2008). *Institutions and organizations—ideas and interests*. London: SAGE.

Sheehan, M. (2007). *The international politics of space*. London: Routledge.

Shepsle, K. A. (1989). Institutional equilibrium and equilibrium institutions. *Journal of Theoretical Politics, 1*(2), 131–147.

Sigalas, E. (2016). Europe in space—The European parliament's justification Arsenal. In T. Hörber & P. Stephenson (Eds.), *European space policy* (pp. 66–81). London and New York: Routledge.

Steinmo, S., Thelen, K., & Longstreth, F. (1992). *Structuring politics: Historical institutionalism in comparative analysis*. Cambridge: Cambridge University Press.

Streeck, W., & Thelen, K. (2005). *Beyond continuity. Institutional change in advanced political economies*. Oxford: OUP.

Suzuki, K. (2003). *Policy logics and institutions of European space collaboration*. Burlington: Ashgate.

Sandholtz, W. & Stone Sweet, A. (1998). *European Integration and Supranational Governance*. Oxford: OUP.

Tallberg, J. (2000). The anatomy of autonomy: An institutional account of variation in supranational influence. *Journal of Common Market Studies, 38*(5), 843–864.

Tallberg, J. (2003). *European governance and supranational institutions: Making states comply*. London and New York: Routledge.

von Kries, W. (2003). Which future for Europe's space agencies? *Space Policy, 19*, 157–161.

Wallace, H. (2000). The policy process. In H. Wallace & W. Wallace (Eds.), *Policy-making in the European union* (pp. 39–64). Oxford: OUP.

Wendt, A. (2001). Driving with the rearview mirror: On the rational science of institutional design. *International Organization, 55*(4), 1019–1049.

Young, O. R. (2010). Regime dynamics: The rise and fall of international regimes. In J. L. Goldstein & R. H. Steinberg (Eds.), *International institutions—Vol. II consequences: When, where and why international institutions are effective* (Vol. II, pp. 163–183). London: SAGE.

Chapter 2
Methodology and Theoretic Framework

While Political Science is often defined as the study of institutions (Steinmo, Thelen, & Longstreth, 1992, p. 3), the study of institutions touches upon several academic disciplines, such as International Relations, International Law, International Political Economy, International History and European Studies (Rittberger & Zangl, 2006, p. 3). The study of the EU seems to be a special case therein, proceeding in waves, initially focusing on explaining the path of integration (i.e. Neofunctionalism and Intergovernmentalism) (Koremenos, Lipson, & Snidal, 2001, p. 766), more recently analysing policy-making with (supranational) institutions and examining the process of institutional change (Caporaso, 1998, p. 2; Meunier & McNamara, 2007, p. 3). Since the mid-1980s institutions became central to the analyses with the rise of New Institutionalism, a diverse research programme, which can be summarised into three sub-approaches, namely Rational Choice Institutionalism, Historical institutionalism and Sociological Institutionalism.[1] Arguably, it is "more a perspective on politics than a fully developed theory" (Peterson & Shackleton, 2006b, p. 5) and there is considerable confusion on how it differs from other approaches (Hall & Taylor, 1996b, p. 5). While arguing that institutions are more than impartial black boxes, its basic lesson is "institutions matter" (Rosamond, 2000, p. 113; Bulmer, 1994; Bulmer & Padgett, 2004). Strikingly however the definitions on "institutions" still vary (Rhodes, Binder, & Rockman, 2006, p. xiii; Mearsheimer, 1994, p. 7) from "organisations", "social norms" (Héritier, 2007, p. 5) and "shared beliefs" to broader understandings such as "the foundation of social life" (Campbell, 2004, p. 1). The term is not only broadly defined (cf. Mearsheimer, 1994, p. 7) but also often used interchangeably with terms such as "organisations" and "regimes". Various schools use different definitions and even within one school different definitions are used (Boin, 2008, p. 89). Given that scholars do not agree on the definition of institutions, they also do not agree on how best to explain or understand the sources of or

[1] The best overview of the defining features, as well as the characteristic strengths and weaknesses of each can be found in Hall and Taylor (1996a, pp. 936–57); for a short overview cf. Rhodes et al. (2006. p. xiii).

N.-L. Remuss, *Theorising Institutional Change: The Impact of the European Integration Process on the Development of Space Activities in Europe*, Springer Theses, https://doi.org/10.1007/978-3-319-95978-8_2

the effects of international institutions (Goldstein & Steinberg, 2010, p. xxiii) and institutional change. The European integration process and the institutions of the EU pose additional challenges on this already diverse existing theoretic framework.

How to explain, both institutional stability and change, is a central concern in Political Science today (Greif & Laitin, 1994, p. 634). While we know a lot about the creation of institutions, relatively little research has been conducted about their transformation once they have been put into place, as has been explained by Héritier (2007, p. 1). The causes and outcomes of institutional change are particularly interesting, as they raise a number of conceptual, theoretical and empirical questions (Héritier, 2007, p. 1). In line with this Shepsle (1989, p. 133) underlined the need to understand the selection, survival, adaptation and evolution of institutional practices. As Campbell (2004) suggests, if we want to specify conditions under which different types of changes are more likely to occur, we need to be able to identify different types of change first (Campbell, 2004, p. 31). In a similar vein Sigalas (2016, p. 77), who analysed the role of the European parliament in framing space policy, recently concluded that there is a lack of a "fully-fledged theory of change". The literature on institutional change seems to lack a structured approach to institutional change. Instead it has dealt with more general questions such as why States rely on international institutions as vehicles of cooperation, what attributes account for their use and how these characteristics set formal institutions apart from regimes and organisations. Moreover, there is a large amount of literature dealing with problems of delegation (cf. Bendor, Glazer, & Hammond, 2000) and principle-agent aspects (cf. Edelmann, Milde, & Weimerskirch, 1998). This is particularly striking as institutional change can be seen as a quasi-permanent process in the EU.

To advance the debate about the impact of the European integration process on existing institutional actors, to make a theoretical contribution to our understanding of institutional change and to make a theoretical contribution to our understanding of when supranational actors are able to advance their own agenda, the aim of this chapter is to provide the theoretical background for the following analysis and develop a methodology. In line with the second movement in the institutional analysis of the 1990s (Campbell, 2004, p. 4), different theories are understood as complementory. In accordance with Keohane and Nye (2001, p. 50) it is analysed, which factors are most important in describing institutional change. The developed methodology proposes a way on how they should be combined. This is in line with Richardson's (1996) observation that the complexity of the EU policy process requires multiple models and the utilisation of "concepts from a range of models in order to at least describe the policy process accurately" (p. 48). Thus, all theoretic accounts under consideration are understood as answering complementary aspects of the same puzzle and the above mentioned theories are analysed in terms of what they offer regarding the analysis of institutional change.

The resulting stages of institutional development as well as the facets of institutional change are proposed as a way to structure the observations of the case of the development of space activities in Europe. Particular attention will be paid to the impact of the European integration process on existing institutional actors and factors working in favour of supranational influence.

First, the main theoretical tenets and their evolution are outlined. Secondly, the importance of the time frame for the analysis of institutional change will be discussed. Third, by relying on Levy, Young, and Zürn (1995) and Florensa (2004), different stages of institutional developments are outlined. Such a differentiation is useful to categorise phases of institutional change and might give insights into the correlation of sources of institutional change and phases of institutional development. Fourth, on the basis of the existing literature mentioned above, five facets of institutional change are identified, which will be used to structure the observations of the case under discussion in the next chapter.

2.1 The State of the Art of the Study of Institutional Change

The study of European integration became a major site for debates in the academic discipline of International Relations, which, by the 1950s, had produced the sub-field of International Organisation (Rosamond, 2000, p. 10). One of the main theoretic accounts analysing the European integration process is Neofunctionalism, explaining European integration as a process of spill-over in which the integration in one economic sector creates pressure for integration of another economic sector. This leads to a deepening of the European integration process (Rosamond, 2000, p. 60; Tallberg, 2003, p. 10). Similarly, concepts such as the "Fusion Thesis" by Wessels (1997) and the concept of "Europeanisation" describe the "ever closer fusion" of the EU.

The second major theory analysing the European integration process is Intergovernmentalism, which developed as a rival to Neofunctionalism. It differs from Neorealism (cf. Buzan, Jones, & Little, 1993) in its recognition of the significance of institutionalisation in international politics and the impact of processes of domestic politics upon governmental preferences. It understands States and national governments as the primary actors in the integration process (Rosamond, 2000, pp. 201–2) and attributes little influence to supranational institutions. Accordingly, institutions frame and guide States (Puchala, 1999, p. 319) and act as "agents of the member States" (Andersen, Eliassen, & Sitter, 2001, p. 28), solving collective action problems (Pierson, 1994, p. 9). These are perceived as inherent in international cooperation. European integration is understood as the result of a series of rational choices made by national leaders, who pursue economic interests (Pollack, 1996, p. 429; Moravcsik, 1998a, p. 319, 326). Any new institutional set-up is explained as either the result of an intergovernmental bargaining process and consequently characterised as the "lowest common denominator" or reflects the interest(s) of the (new) hegemon.[2] If agreement on a new institution cannot be reached, the institution is said to cease to exist.

[2]Krasner argues that when States have different preferences over the possible cooperative outcomes that can be achieved under different institutional forms, the power of the States involved will determine the outcome achieved (cf. Conner, 2010, p. iix).

Liberal Intergovernmentalism (LI) was developed by Moravcsik (1993a) as a variant of Intergovernmentalism, introducing the two-stage preference formation in intergovernmental bargaining. Accordingly, in the first stage, national preferences are determined by constraints and opportunities imposed by economic interdependence, while in the second stage, the outcomes of intergovernmental negotiations are determined by the relative bargaining power of governments (Moravcsik, 1993a, p. 517).

An alternative theoretic approach to LI is Supranational Governance (cf. Marks et al., 1998; Pierson, 1996; Sandholtz & Stone Sweet, 1998; Jachtenfuchs, 2001; Kohler-Koch, 2000). While there are considerable variations among the ideas of the proponents, they share the common claim of significant independent influence of supranational institutions due to the lack of full control by the Member States (Hix, 2005, p. 16).

Moreover, with the increasing number of agencies being founded within the EU institutional framework an increasing number of authors[3] started dealing with the phenomenon "agencification", i.e. the increase in the numbers of community agencies (cf. Christensen & Nielsen, 2010). Additionally, Framing Theory has become increasingly popular in recent European Studies with authors identifying frames operating in the discourse of different actors (cf. Kohler-Koch, 2000; Daviter, 2007; Hörber & Stephenson, 2016).

A groundswell of academic momentum developed behind the idea that institutions were important but neglected and it was time to bring them "back in" to the study of politics in the mid-1980s, giving rise to New Institutionalism.[4] For some time, there was considerable confusion about what "New Institutionalism" is (Hall & Taylor, 1996a, p. 936), as is proven by the numerous essays, examining the institutional and historical turn of political sciences. Even today no single orientation characterises the vast scholarship falling under this heading (Rhodes et al., 2006, p. xiii). Rather the literature is diverse and does not constitute a single research programme (Ibid.; Rosamond, 2000, p. 113), but at least three different analytical approaches (Hall & Taylor, 1996a, p. 936): Rational Choice Institutionalism, Historical Institutionalism, and Sociological Institutionalism.[5] Each of these schools is characterised by "tremendous internal diversity" (Thelen, 1999, p. 370) and it is often difficult to distinguish sharply between the three schools. It is said to be more a perspective on

[3]e.g. Trondal and Jeppesen (2008), Yataganas (2001), Geradin and Petit (2004), Wonka and Rittberger (2010), Pollack (1997b), Rittberger and Wonka (2011), Majone (2001), Egeberg and Trondal (2010), DiMaggio (1988).

[4]Also, the internal market programme, launched in 1985, had given new political impetus to the stagnating integration process and served as a stimulus for the question of classical integration theory and a renewed debate on how the development of the European integration process could be explained. The turning point is also often said to be provided by the seminar article on the joint-decision trap Cf. as well as by the influential article by March and Olsen (1984), calling for the rediscovery of institutional thinking in political science and heralded the institutionalist turn in political science. Cf. Peterson and Shackleton (2006b, p. 5), Jachtenfuchs (2001, p. 249), Boin (2008, p. 88), Aspinwall (2000).

[5]The best overview of the defining features, as well as the characteristic strengths and weaknesses, of each can be found in Hall and Taylor (1996a, pp. 936–57).

politics than a fully developed theory (Peterson & Shackleton, 2006b, p. 5). There is considerable confusion how it differs from other approaches (Hall & Taylor, 1996b, p. 5). Still, New Institutionalism has emerged as a leading, even arguably dominant perspective on European integration and politics (Peterson & Shackleton, 2006b, p. 5). The degree to which institutions matter (Puchala, 1999, p. 318), is central to its analysis. Its core concepts are institutions, institutionalisation and institutional change (Sjöstrand, 1993, p. 3). Its basic lesson is "institutions matter" (Rosamond, 2000, p. 113; Bulmer, 1994), as they are "the source of much political behaviour and not impartial "black boxes", which simply transform preferences into policies." (Peterson & Bomberg, 1999, p. 16). Institutions "make a difference" (Puchala, 1999, p. 318) by helping to organise political activity, by creating and moulding identities, defining what is considered successful or legitimate, and giving authority to certain actors and not others. Institutions are understood to "provide the means of change when actors desire it, or offer continuity." (Warleigh, 2001, p. 7). If institutions matter, then which institutions are employed becomes an important question. Particularly, the selection, survival, adaptation and evolution of institutional practices need to be understood (Shepsle, 1989a, p. 133). Institutions are perceived to solve problems of coordination among States (Lake, 2001, p. 130). They are the "rules" of the game and organisations are "the players" (Peterson & Bomberg, 1999, p. 17).

Among the disciplines mentioned, above the question about the impact of the European integration process on existing institutional actors seems to be a case, which has so far not been covered substantially. While several authors tried to conceptualise the relationship between the EU and other intergovernmental organisations (in particular NATO, Council of Europe, OSCE), no structured debate can be identified. Instead the existing literature on institutions has dealt with more general questions such as why States rely on international institutions as vehicles of cooperation, what attributes account for their use, and how these characteristics set formal institutions apart from regimes and organisations. Moreover, there is a great amount of literature dealing with problems of delegation (cf. Bendor et al., 2000) and principle-agent aspects. At the same time the literature on institutional change lacks a structured approach, describing which factors are most important in institutional change (Keohane & Nye, 2001, p. 50).

Theories are valuable "because they structure all observations" (Rosamond, 2000, p. 4). Beginning in the late 1990s scholars began calling for a second movement in institutional analysis and asked for "a more constructive dialogue" (Campbell, 2004, p. 4), exploring the ways in which different theories "complement and connect to each other" (Campbell, 2004, p. 4). It was argued that no single model is likely adequate to explain world politics as conditions vary too greatly (Keohane & Nye, 2001, p. 50). Most scholars agree that no single theory can account adequately for everything (Egeberg, 2004, p. 199). In the same vein, it is "unlikely that one theory can claim to account for institutional change under each and every circumstance" (Héritier, 2007, p. 39). Additionally, "[i]t is not obvious that any one approach is superior to the others in capturing the complexities of change. There are several stories to be told (…)." (March & Olsen, 1998, p. 964). Rather each theory has "a selective grasp on reality" (Héritier, 2007, p. 39). Scharpf (1997) for example understood the combination of different theories as a necessity to get a "more complete explanation" (Scharpf, 1997, p. 31). Keohane and Nye (2001, p. 50) suggested to analyse which factors are most important and how they should be combined. In accordance with this a more constructive dialogue between different theories and paradigms is suggested by this dissertation. It is argued, they should be considered as complementary lenses.

Most strikingly, by taking stock of the existing theoretic accounts on institutions, no singular definition for the term "institution" could be identified (Rhodes et al., 2006. p. xiii; Mearsheimer, 1994, p. 7). As Mearsheimer (1994, p. 7) observes, "[t]he concept is sometimes defined so broadly as to encompass all of international relations." The above-mentioned theoretic evolution is also mirrored by the changing terms used in the literature ranging from "institutions" to "organisations" to "regimes", which are sometimes even used interchangeably. Various schools rely on different definitions and even within one school, different definitions are used (Boin, 2008, p. 89), ranging from institutions as "organisations", "social norms" (Héritier, 2007, p. 5) and "shared beliefs" to broader understandings such as "the foundation of social life" (Campbell, 2004, p. 1). Institutions are also described as "formal, demographic, physical, temporal, cognitive and symbolic arrangements" (Egeberg, 1994, p. 85). Even though the term "international organisation" is often used synonymous to the term "institution", international organisations can in fact be understood as a specific class of international institutions (Rittberger & Zangl, 2006, p. 6), characterised by a secretariat or other agent, often enjoying a permanent physical headquarter. Some are intergovernmental, some non-governmental. They may have a single, narrow purpose, such as advancing the protection of intellectual property rights, or multiple broad purposes (Goldstein & Steinberg, 2010) and vary in their competencies. Rittberger and Zangl (2006, pp. 6–7) further distinguish between "international regimes" and "international organisations". International regimes are according to (Barnett & Finnemore, 1999) issue-specific international institutions, which do not function as actors but rather as "principles, norms, rules and decision-making procedures" (p. 700). This study defines "institutions" as organisations and is thus not considering them as norms, rules or any other types of regimes.

Given that scholars do not agree on the definition of institutions, they consequently also do not agree on how best to explain or understand either the source or the effect of international institutions (Goldstein & Steinberg, 2010, p. xxiii). Héritier (2007) has advanced the debate on institutional change by asking: Which are the major rational institutionalist theories accounting for institutional change? What are the underlying argument accounting for change and under which particular conditions are they particularly apt to explain a particular type of institutional change? (Héritier, 2007, p. 9).

I opt for a set of different questions:

- What are the sources for institutional change?
- Which factors are most important in describing institutional change?
- Which stages of institutional development can be identified?
- Which outcomes of institutional change are possible?
- How can different cases of institutional change best be compared?

- What is the impact of the European integration process on existing institutional actors? In what situations can governments be expected to dominate policy-making and in what supranational actors?

In this analysis institutions are thus understood to be both, the independent and dependent variable, both influencing actors in their choices and being shaped by actors.

2.2 The Importance of the Time Frame for the Analysis of Institutional Change

Several authors (e.g. Aminzade 1992; Campbell 2004) suggest paying attention to the time frame, over which institutional change is examined. Accordingly, the appropriate time frame for analysis depends on the historical specificity of the phenomenon in question "and cannot be prescribed a priori for all social phenomena" (Campbell, 2004, p. 44), since different social processes have different rhythms. Campbell (2004) suggests, selecting the time frame on the basis of a consideration of the phenomenon and its historical context. Critical events may mark the appropriate time frames for analysis (e.g. electoral, budget, legislative or other cycles). Aminzade (1992) highlights that temporal concepts allow to situate events in a context of relevant larger-scale and longer-term processes, thereby being able to identify potential sources of institutional change. Accordingly, the sequencing of events is important in determining how institutional change occurs. This way, sequences of events "that have unfolded in similar but not identical fashion in a variety of different historical contexts" (Aminzade, 1992, p. 458) can be analysed. Temporal concepts help to understand how events fit together in sequences.

Surprisingly, institutionalists have so far neglected the issue of how to specify the time frame over which institutions are tracked in an effort to determine how much institutional change has occurred. The lack of attention to the issue of time frame may, according to Campbell (2004, pp. 41–3), result in empirical findings that incorrectly specify the boundaries, turning points, or critical events that separate different institutional moments.

2.3 Stages of Institutional Development

According to Sandholtz & Stone Sweet (1998), institutions are in constant evolution, rarely emerge at once or "change as a single unit" (p. 16). What we perceive as institutional change is presented as the "normal" process of institutional evolution. In an effort to identify different types of change and factors most important to describing institutional change, one can distinguish different stages of institutional development on the basis of Levy et al. (1995) and Florensa (2004). Combining their suggestions, the process of institutional evolution seems to consist of five stages, which proceed continuously:

- agenda formation (Levy et al., 1995),
- institutional choice (i.e. selection and design) (Ibid.),
- operationalisation (Ibid.) or evolution (Florensa, 2004)
- redefinition (Ibid.) and
- crisis (Ibid.).

These five stages are proposed as an element to structure the observations of institutional change and will be applied to the case study. They are further detailed below.

2.3.1 Stage 1—Agenda Formation

The first stage of institutional development as suggested by Levy et al., (1995) is the stage of agenda formation. It comprises the emergence of a topic on the political agenda, the framing of the issue for consideration in different forums and "the rise of the issue to a high enough place on the international agenda to the point of agreement" (Levy et al., 1995, p. 282). This stage is particularly interesting since it gives insights into how the question to found or change an institution is put on the agenda. The critical question in this stage is how an issue or demand is considered important or not important to be included on the agenda (Cobb & Elder, 1971, pp. 903–4), and more specifically, why certain issues end up as topics for the EU (Princen, 2007). In line with Cobb and Elder (1971), the "agenda" is defined as a "general set of political controversies (…) meriting the attention of the polity" (Cobb & Elder, 1971, p. 905). Accordingly, while the composition of agendas varies of periods and times, there is a tendency of decision-makers to give priority to topics, which have previously reached the agenda but have either not been resolved or are subject to periodic alteration (Cobb & Elder, 1971, pp. 906–7).

According to Princen (2007) it is impossible to identify the ultimate source of an idea and thus more fruitful to look at a set of factors accounting for the rise and fall of issues on the political agenda. Instead he proposes to focus on the process in which actors form coalitions and issues are framed and on the conditions under which actors succeed in getting issues on the agenda (Princen, 2007, p. 23). In Neorealist thinking (cf. Waltz, 1979), States' worry about their survival is the main driving force (Princen, 2007, p. 24). According to Regime Theory collective action problems such as technical standardisation form the condition for issues to get on the agenda (Princen, 2007, p. 25). According to Princen (2011, p. 929) the two key elements in agenda-setting are decision-making and issue frames. While for a long time the study of agenda-setting hat been dominated by established parts of Political Science and policy analysis, issue frames have been found to influence agenda-setting more recently. According to Daviter (2007, p. 655) two theoretical perspectives can be distinguished. The first goes back to Schattschneider's (1957, 1960) conception of politics "according to which political conflict and competition structure (…) fix the attention of the public, shape the organization of interest and the formation of coalitions and alliances" (cf. Daviter 2007, p. 656). The second goes back to Rein and Schön (1991), who focused on social processes to resolve substantial conflicts of interest (Ibid., p. 657). They define framing as "a way of selecting, organising, interpreting, and making sense of a complex reality so as to provide guideposts for knowing, analysing, persuading, and acting" (Rein & Schön, 1991, p. 263), leading to different views of the world and creating multiple social realities (Ibid., p. 264). According to Kohler-Koch (2000, p. 516) framing "colours the nature of options".

In the context of agenda formation, Rittberger and Zangl (2006, p. 25) have identified three conditions, which they perceived as essential for the foundation of an institution:

- the "problem condition", i.e. complex interdependencies, which push States into international cooperation to further common interests;
- the "cognitive condition", i.e. the common interest that can only be achieved through cooperation within an international organisation and
- the "hegemonic condition", i.e. a hegemonic State willing to bear the costs of the creation of an institution.

Similarly, according to DiMaggio (1988, p. 14) new institutions arise, when organised actors, so-called institutional entrepreneurs (cf. Fligstein, 2010), with sufficient resources (e.g. "hegemonic condition") see in them an opportunity to realise their interests ("problem condition"). Olsen (1997, p. 209–10, 217) adds two more possibilities. Accordingly, institutions can be created by exploiting "windows of opportunity" at formative events (Ibid., Mazey & Richardson, 1996, p. 42) or breaking points, depending on the ability to exploit shifts in the general ideological climate or Zeitgeist, as well as sudden surprises or major external shocks. Issues are said to not be immune to external events, which can lead to changes in the perceptions of policy issues and in a reshuffling of policy positions (Daviter, 2012, p. 12).

Considering the European integration process, the phase of the agenda-formation is an important determinant for the policy outcomes. At the same time, EU involvement for many issues is not self-evident (Princen, 2011, p. 930). First, there is the formal aspect of legal competence, given the principle of subsidiarity and the operation under a system of conferred competencies. Secondly, the EU needs to have sufficient expertise and organisational capabilities. Third, the EU is only one among a number of international venues to deal with a certain topic (Princen, 2011, p. 930). For issues where EU involvement is not self-evident, it is not sufficient to convince policy-makers of the significance of the issue but it is also necessary to convince them, that the EU should take it up. Prospective agenda-setters need to frame an issue in a way that it arouses interest by tying it with established overall-values, central to the EU's purpose, priorities and identity such as the common market. Alternatively, small steps, such as doing a study, organising a conference or focusing on specific non-controversial elements of target groups, can help for an issue to arouse interest (Princen, 2011, pp. 930–4). When the issue made it to the EU agenda, proponents need to build up sufficient organisational capacity to build credibility and to construct a story about why an issue is European in scope, by linking it to EU causes and competences (Princen, 2011, pp. 934–7). During the agenda-setting phase, the monopoly of legislative initiative lies with the EU's Commission (Hartlapp, Metz, & Rauh, 2013, p. 425), which is often said act as a "purposeful opportunist" (Cram, 1993, p. 136) aiming at expanding its influence.

2.3.2 Stage 2—Institutional Selection, -Choice and -Design

According to Levy et al. (1995), the stage of institutional selection refers to "the point of agreement on the provisions of a specific regime." (p. 82). Several authors (e.g. Thelen 1999, p. 400; Hall & Taylor 1996a; Connor 2010, p. iiix) link the process of institutional foundation and selection to the process of institutional change. Accordingly, knowing how institutions were constructed "provides insights into how they might come apart" (Thelen, 1999, p. 400) and "institutions survive primarily because they provide more benefits to the relevant actors than alternate institutional forms" (Hall & Taylor, 1996a, p. 945). Connor (2010, p. iiix) and Hall & Taylor (1996a) suggest that institutions are doomed to fail when they are not successful "in carrying out the goals attendant with their creation?" (Hall & Taylor, 1996a, p. 952). Thus, institutions rest on different sets of "ideational and material foundations" (Thelen, 1999, p. 369). Given these different foundations, the processes that are likely to change them, will also be different (Thelen, 1999, p. 369). Analysing the institutional design, will give insights into different institutional foundations. This way institutional change and the way the institution will change, will become more predictable (Steinmo et al., 1992, p. 27).

Pollack (1996, p. 441), Koremenos et al. (2001) and Hall (1996) suggest "institutions are chosen" (Pollack, 1996, p. 433) or competitively selected by "States (…), other international actors" (Koremenos et al., 2001, p. 767) and "relevant actors between different alternating institutional forms or designs" (Hall & Taylor, 1996a, p. 945) and do not simply happen (Pollack, 1996, p. 433). According to Cobb and Elder (1971), the "pre-political, or at least pre-decisional, processes are often of the most critical importance in determining which issues and alternatives are to be considered (…) and which choices will probably be made" (p. 903).

There seem to be three competing arguments about the reasons for the selection of one particular institution. First, according to proponents of "rational design" (Wendt, 2001, pp. 1023–4) such as Intergovernmentalists and Liberal Intergovernmentalists (cf. Dehousse & Magnette, 2006; Pollack, 1996, p. 441, 2001; Moravcsik, 1993a), institutions are selected on the basis of rational expectations regarding the costs and benefits expected (Pollack, 1996, p. 434; Wendt, 2001, p. 1043) through a process of intergovernmental bargaining on the basis of national (mainly economic) preferences (Dehousse & Magnette, 2006, p. 29) and other "motivational factors" (Wallace, 2000b, pp. 43–4). The outcomes of intergovernmental negotiations are suggested to be dependent on the relative bargaining power of governments (Moravcsik, 1993a, p. 517). Some authors such as Héritier (2007, p. 12), Williamson (1985) and Pierson

(2000, p. 476) further distinguish the functional intentional approach by including and excluding transaction costs. Accordingly, the development of a particular organisational form can be explained as the result of the efforts of rational actors to reduce transaction costs.[6] Secondly, according to Wendt (2001, p. 1023), institutions are selected on the basis of what is normatively appropriate. Third, as Héritier (2007, p. 1) observes, institutions can also be selected unintentionally without grand intentional decisions. This happens when informal institutional rules persist over a longer period of time. As Jupille (2004, p. 1) observes, the availability of institutional alternatives constrains actors. In line with Scott (2008, p. 94), institutions do not emerge in a vacuum but always challenge, borrow from and to varying degrees display prior institutions.

Additionally, three different factors seem to influence the institutional selection, i.e.

- contextual factors, derived from the broad circumstances;
- functional factors, derived from some of the core functions of politics and demands to be met, and
- institutional factors, i.e. opportunities and constraints which influence in how far institutions are able to channel and to structure the behaviour of political actors from the participating countries (Daviter, 2012, p. 9; Wallace, 2000b, pp. 43–4).

2.3.3 *Stage 3—Operationalisation*

According to Levy et al. (1995), the third stage of institutional development is operationalisation. It "covers all those activities required to transform an agreement on paper into a functioning social practice" (p. 282). Florensa (2004) refers to this stage as "evolution", which she defines as "learning processes, by the acquisition of new skills, information and knowledge by the actors involved, leading to readjustments in the institutional structure" (p. 9).

2.3.4 *Stage 4—Redefinition*

Suggesting that institutions follow a life-cycle composed of periods of stability and change, Florensa (2004) describes the fourth stage of institutional developments as "redefinition", i.e. institutional change consisting "of marginal changes of the institutional structure" (p. 9).

[6] Institutions are understood to build up expertise and infrastructure. These are summarised as transaction costs. When a given institution is established the transaction costs of changing that institution are not negligible.

2.3.5 Stage 5—Crisis

Florensa (2004) describes the fifth stage as crisis, defining it as "comprehensive institutional change" (p. 9). Accordingly (p. 10), once change has taken place and crises are past, institutions are likely to tend towards stability again through self-reproduction and reinforcement.

2.4 Facets of Institutional Change

One of the objectives of this dissertation is to make a theoretical contribution to our understanding of institutional change. To do so existing theoretical accounts dealing with institutional change have been analysed in terms of which factors are most important in describing institutional change, as called for by Keohane and Nye (2001, p. 50). The aim is to identify different types of institutional change as called for by Campbell (2004, p. 31). As a result, a set of facets[7] of institutional change is being proposed, describing institutional change and making different cases of institutional change comparable. This offers a more structured approach to the analysis of institutional change:

- the degree of change (incremental or revolutionary change; changes in different institutional dimensions),
- the pace of institutional change (fast, slow),
- sources of institutional change (endogenous, exogenous),
- change agents and
- forms of institutional change.

2.4.1 Degree of Institutional Change

The first facet of institutional change, which can be identified, is the degree of change. Hall (1993, p. 278) differentiates between simple change, i.e. in instruments and means, and complex change, i.e. in goals and ends. Florensa (2004, p. 9) offers a more differentiated analysis. Accordingly, change can take place at different levels of the institutional structure and change in one level again can lead to changes at deeper levels of the institutional structure or the reverse. Similarly, Campbell (2004) suggests distinguishing the degree of change by qualifying in which institutional dimension it took place. In line with Hall (1993) and Campbell (2004), the following differentiation regarding the degree of changes is suggested

- simple change, i.e. in instruments and means (Hall, 1993, p. 278),

[7] A facet is understood as "one of several parts" (cf. Longman Dictionary of Contemporary English) being synonymous to "aspects".

- complex change, i.e. change in the normative[8] institutional dimension,
- change in the regulative[9] institutional dimensions of an institution (Campbell, 2004, p. 40), and
- change in the cultural cognitive[10] institutional dimension (Ibid., p. 36; p. 48).

In line with this, Campbell (2004, p. 40) offers, if we want to determine, how much institutional change has occurred in a particular setting, we need to examine over time all the relevant institutional dimensions.

2.4.2 Pace of Institutional Change

The second facet of institutional change is pace. Pace is closely related to the degree of change, as it can also be measured by the amount of institutional dimensions subject to change over a given period of time (Campbell, 2004, p. 71).

Several authors (i.e. Campbell, 2004; Madariaga, 2011; Roland 2004) seem to agree on two types of paces of institutional change, i.e.

- slow, incremental through small adjustments, gradual or evolutionary in a continuous manner and
- rapid revolutionary change through large steps in a discontinuous manner, resulting in a "steep transformation entailing a breakdown and replacement of old institutional arrangements" (Madariaga, 2011, p. 17). This is also referred to as "path cessation" or "path switching" (Ebbinghaus, 2005, p. 17).

In line with this, Hall (1993) argues that adjustments in both instruments and settings occur incrementally, while changes in goals are rare and more fundamental (in Campbell, 2004, p. 48).

Many Historical Institutionalists (i.e. North, 1990, p. 73; March & Olsen, 1989; Pollack, 1996, p. 441; Campbell, 2004) consider incremental change as typical, referring to it as the "main story" (March & Olsen, 1989, p. 169), with institutional stability and continuity being interrupted by "occasional periods of rapid change" (March & Olsen, 1989, p. 169) or "punctuated equilibrium" (Campbell, 2004, p. 34; Pollack, 1996, p. 439; North, 1990, p. 6; Baumgartner & Jones, 1993). Additionally, Incremental small changes extending over a long period of time can in their sum amount to substantial institutional change (Héritier, 2007, p. 1; Thelen, 2003; Pierson, 2000).

---

[8]e.g. principles prescribing goals of behaviours and appropriate ways to pursue them.

[9]e.g. legal, constitutional or other rules that constrain and regulate behaviour.

[10]e.g. culturally shaped assumptions about reality and the frames through which it is perceived, understood and given meaning.

2.4.3 *Sources of Institutional Change*

The literature distinguishes between endogenous (i.e. Greif & Laitin, 1994; Streeck & Thelen, 2005; Young 2010) and exogenous (Pollack, 1996, p. 441; Krasner, 1988, p. 67; Scott, 2008; Steinmo et al., 1992; Suzuki, 2003) sources of institutional change.

The main proponents of endogenous sources of institutional change are Greif & Laitin (1994), who propose a theory of endogenous institutional change. Accordingly, endogenous factors can lead to a self-undermining process, which once it "reaches a critical level" (p. 634) results in institutional change. In a similar vein, Streeck and Thelen (2005) highlight endogenous sources of institutional change, which result from "inherent ambiguities and "gaps" that exist by design or emerge over time" (p. 19). They perceive institutional design as one potential endogenous source of institutional change, highlighting unfulfilled expectation as a source, leading to questions of form and function of the institution, with one possible outcome being institutional change (Ibid.). Similarly, Young (2010) refers to "internal contradictions" resulting from institutional design, which might lead "to serious failures mounting pressure for major alterations" (p. 174). In the same vein, Héritier (2007) refers to "institutional properties", setting "a causal process into motion, resulting in altered institutional properties without outside influence" (p. 8). Florensa (2004, p. 6) refers to ideological innovation, learning processes through the acquisition of knowledge as well as skills and changes in tastes or preferences of the actors involved, as examples for endogenous sources for institutional change.

Looking at the existing literature examples for exogenous sources for institutional change are:

- "critical junctures or punctuation" (cf. Historical Institutionalism)[11] also referred to as "windows of opportunities" (i.e. changes in the attribution of natural resources, extreme climatic events or natural disasters),
- changing preferences of supportive or influencing governments (Florensa, 2004, p. 18; Pollack, 1996, p. 441), which might result from unfulfilled expectations,
- shifts in power distributions of previously supportive coalitions or the appearance of new actors (Pollack, 1996, p. 438; Scott, 2008, p. 198),
- changing voter preferences, leading to new political alignments,
- lobby of business interest (Scott, 2008, p. 196),
- broad changes in the policy environment of social, political or economic nature (Steinmo et al., 1992, p. 16),[12] and
- improvements in the quality of information about either the operation of the institution or the environment within which it operates (Pollack, 1996, p. 438).

[11] For more information on "critical junctures" refer to the following literature cf. Florensa (2004, p. 18), Christiansen and Vanhoonacker (2008), Campbell (2004, p. 26), Hall and Taylor (1996a, p. 942), Krasner (1984, p. 235 in Ikenberry, 1994, p. 9), Horak (2007, p. 21), March and Olsen (2005, p. 12).

[12] For example, changes in the markets, technological innovations or legislative changes (cf. Florensa 2004, p. 18; Pollack 1996, p. 438).

According to Snidal (2010), any changes in the circumstances of context of an issue area can produce institutional change (Snidal, 2010). This is in line with Steinmo et al. (1992) model of "institutional dynamics", which focuses on the interaction between institutions and political processes. Accordingly, changes in socio-economic and political contexts can influence existing institutions. Similarly, Florensa (2004, p. 9) refers to "surrounding social, economic or political context" as sources of institutional change. None of the authors however suggests, under what conditions, which sources preside. May be our case study will give further insights.

2.4.4 Change Agents

According to Madariaga (2011) sources of institutional change "have to be activated by agents willing to generate change" (p. 24), such as States (and their ministries, representatives), individuals, groups (interest groups and communities of experts) or institutions (administrative staff of international organisations; parliamentary assemblies of international organisation) (Rittberger & Zangl, 2006, p. 78). In this context the question arises, under what conditions each of the change agents is able to advance its own agenda. In line with Héritier (2007, p. 8) a structural perspective on institutional change is adopted by looking at the types of actors involved in the process of change.

2.4.4.1 States

The first so-called change agent identified by Madariaga (2011) are States. According to theoretic accounts such as Realism/Neo-realism (Rosamond, 2000, p. 76), Intergovernmentalism/LI and Neoliberalism (Rittberger & Zangl, 2006, p. 18), Historical Institutionalism and Rational Choice Institutionalism States are the main actors creating, sustaining and changing institutions on the basis of rational calculation (Suzuki, 2003, p. 2). The existing literature gives several examples of reasons for States to act as change agents, triggering a process of change. Those are:

- changes in State preferences (Pollack, 1996, p. 441), changes in preferences of supportive governments (Ibid.), (which might result from unfulfilled expectations)
- shifts in power distributions of previously supportive coalitions or the appearance of new actors or a new hegemon (Florensa, 2004, p. 18; Pollack, 1996, p. 438; Scott, 2008, p. 198),
- changing voter preferences, leading to new political alignments,
- lobby of business interest (Scott, 2008, p. 196),
- broad changes in the policy environment of social, political or economic nature (Steinmo et al., 1992, p. 16),
- changes in the original collective action problems,
- changes in the circumstances or context of an issue area (Snidal, 2010),

- contract corrections, and
- improvements in the quality of information about either the operation of the institution or the environment within which it operates (Pollack, 1996, p. 438).

According to Realism/Neo-realism (Rosamond, 2000, p. 76), Intergovernmentalism/LI and Neoliberalism (Rittberger & Zangl, 2006, p. 18), Historical Institutionalism and Rational Choice Institutionalism, institutions act as "agents of the member States" (Andersen et al., 2001, p. 28), solving collective action problems (Pierson, 1994, p. 9) inherent in international cooperation. European integration is thus understood as the result of a series of rational choices made by national leaders, who pursue economic interests (Pollack, 1996, p. 429; Moravcsik, 1998a, p. 319, 326). Any new institutional set-up will thus either be the result of an intergovernmental bargaining process or reflect the interest(s) of a (new) hegemon.[13] If agreement on a new institutional rule cannot be reached, the institution is said to cease to exist. Neorealists and Rational Choice Institutionalists reason that institutions change, when State preferences change (Héritier, 2007, p. 12; Suzuki, 2003, p. 2, 31), for example due to economic reasons, changes in the balance of power between the involved actors, changes in the bargaining power of the actors, in the geopolitical interests (Moravcsik, 1998a, p. 27), in ideology (Ibid.) or technological changes in the environment (Héritier, 2007, p. 42). Thereby, the implementation of national preferences is said to be limited by the international and domestic context (Suzuki, 2003, p. 4), the EU framework and the State's power, i.e. its position in the international system (Héritier, 2007, p. 42). Neorealism, Intergovernmentalism and Regime Theory allocate a special role to the most powerful States in the system (Rosamond, 2000, p. 113) or the balance of power of the cooperating Sates. They allocate the source of institutional change in changes of the balance of power.

Regime Theory postulates the decline of economic hegemons as the source of regime change by drawing on the insights of Transaction Cost Economics (Pierson, 1998, p. 32). Once a hegemon declines, institutions will be adapted as to reflect the interest of the new hegemon or the increasingly powerful States or entities (Goldstein & Steinberg, 2010, p. xxi).

Rational Choice Institutionalists also reason institutions change or are replaced when they cannot or can no longer provide benefits (Suzuki, 2003, p. 9), which can be a result of preference shifts and increases in transactional costs. Institutional change becomes possible once the bargaining power of a State changes. Governments negotiate on the basis of national preferences and motivational factors when selecting an institution. Since, the outcomes of intergovernmental negotiations are dependent on the relative bargaining power of governments (Moravcsik, 1993a, p. 517), the degree to which States can influence institutional change also depends on their bargaining power.

Neoliberalism explains the creation of institutions as a tool for States to cooperate successfully in the pursuit of a common interest when these interests are neither

---

[13] Krasner (2010) argues that when States have different preferences over the possible cooperative outcomes that can be achieved under different institutional forms, the power of the States involved will determine the outcome achieved (cf. Conner, 2010, p. iix).

totally aligned nor mutually exclusive (Rittberger & Zangl, 2006, p. 18). Proponents of Rational Choice Institutionalism argue that European countries collaborate because they want to reduce risks and cost of research and development, thereby strengthening national industry and increasing financial support (Suzuki, 2003, p. 10). If institutions are founded to solve a problem of collaboration and coordination as proposed by Abbott and Snidal (2010, p. 263), then States will change or dissolve institutions once this collective action problem changes (Héritier, 2007, p. 15). This can be the result of changes in the external environment,[14] making a new institutional solution seem more favourable. States will weigh the transaction costs caused by the redesigning of the existing institution (i.e. seeking information, engaging in a bargaining process etc.) against the costs of keeping the institution as it is (Héritier, 2007, p. 16).

In line with the principal agent approach, institutions through some kind of contract act as the agents of States (Yi-Chong & Weller, 2008, p. 36). States can thus act as change agents once they perceive the need for contract correction, resulting from for example extensive agency loss (Tallberg, 2003, p. 5).

Historically, governments have played a major role in initiating and funding space activities (Aliberti & Krasner, 2016; Brachet, 2016). The field of space is particularly prone to State influence, as national space programmes and the European space programme both conducted by a national space agency funded by the government are the rules of the game. The sector of satellite-based telecommunication is the only one where the commercial market started playing a role as early as the 1980s. Thus, the sector developed independently of government funding with governments continuing to play a strong regulatory role (Brachet, 2016, p. 211). Launch services recently gained a relative autonomy with respect of government technical control with private companies such as Space-X and Blue Origin demonstrating to provide a reliable service (Brachet, 2016, p. 211). According to a survey ESA Member States prioritise technology, sustainability and motivation (Giannopapa, Adriaensen, & Sagath, 2016, p. 188).

2.4.4.2 Individuals

The second so-called change agent identified by Madariaga (2011) is individuals, i.e. chief negotiators, institutional entrepreneurs, or experts. The main proponents of such an approach are Fligstein (2010), North (1990, p. 73), Sabatier and Jenkins-Smith (1996, cf. Sabatier, 1998, p. 109), who consider so-called institutional or policy entrepreneurs, defined as "individual and organisational actors, who in line with their own preferences create opportunities for innovation and institutional change",[15] as change agents. Rational Choice Institutionalists such as Campbell (2004) also

[14] In that sense, this approach is quite similar to the phenomenon of critical junctures proposed by Historical Institutionalism.

[15] Institutional entrepreneurs are individual and organisational actors, who in line with their own preferences create opportunities for innovation and institutional change. (cf. Fligstein, 2010).

analysed, how individuals build and modify institutions to achieve their interest (Campbell, 2004, p. 15). Additionally, Putnam (1988, p. 434ff) examined the role of the chief negotiator, whose preferences can diverge from those of the domestic principals, and who may be able to influence an agreement through the dual role at both the international and the domestic bargaining tables. In line with Cobb and Elder (1971) for an issue to attain agenda status, it must "command support of at least some key decision makers" (p. 907), as they are the ultimate guardians of agenda.

The existing literature gives several examples of reasons for individuals to act as change agents, triggering a process of change:

- unfulfilled expectations,
- changes in interests,
- excessive maintenance costs (Florensa, 2004, p. 8),
- windows of opportunities.[16]

According to Copeland and James (2014, p. 3) there are several ways for individuals to react to windows of opportunities. Constituting information signals from the broader policy environment, intentional agents respond to them with "coupling". Accordingly, for policy change (i.e. institutional change) to occur, policy entrepreneurs couple different issues to construct a narrative that mobilises political action around a policy problem to legitimise a certain solution.

Ackrill and Kay (2011, p. 75) further distinguish between policy entrepreneurs, i.e. individuals that sell policies to policy makers, and policy commissioners, i.e. individuals who select policies.

According to Fligstein (1997a, b), the degree to which individuals are successful in acting as a change agent is dependent on their ability to frame "stories"[17] and their "social capital", i.e. the acceptance of the actors. According to Stephenson (2016, pp. 31–2), policy makers construct policy problems as well as they offer policy solutions by framing issues. They act as designers, engaging in dialogue with experts and interests, and play an active role in the construction of the images and venues used to communicate issues. Frames alter perceptions, beliefs and appreciation, determining how people understand an issue and what counts as facts for them. Actors sponsor frames according to their interests. Actors' frames are shaped by their backgrounds (Giannopapa et al., 2016, p. 173). According to Moravcsik (1999, p. 275), the most common explanation for the influence of individuals is that certain international officials are more ingenious, imaginative, skilful, and creative than national leaders. They thus wield power by proposing creative solutions to political problems.

Early studies on the role of the Commission in the integration process according to Kassim and Dimitrakopolous (2007) "emphasise the importance of the Commission President's conviction concerning the development of the Union and his leadership style as key factors" (p. 1252) shaping the Commission's preferences. Often Prodi,

[16]Windows of opportunities constitute information signals from the broader policy environment (Copeland & James, 2014, p. 3).

[17]With the objective to induce cooperation from people that appeal to their identity and interests, while at the same time using those same stories to frame actions against various opponents (Fligstein, 1997b, p. 17).

who underlined the Commission's centrality to the European project (Ibid., p. 1253), is considered a prime example. The same research also highlights the role of officials at various levels of administration (Ibid., p. 1252).

2.4.4.3 Groups

The third so-called change agent identified by Madariaga (2011) is groups, such as expert groups, epistemic communities, policy networks and advocacy coalitions. There are several competing theoretical attempts theorising the influence of groups in policy making: (1) Haas' (1992, 2010) concept of epistemic communities, (2) Sabatier's concept of advocacy coalitions (Sabatier & Jenkins-Smith, 1996)[18], (3) the policy networks approach (Börzel, 1998) as well as concepts such as (4) "brusselisation", "socialisation" and "bureaucratisation" (cf. Duke & Vanhoonacker, 2006), which can be applied to institutional change. The distinction between them has led to an "overly desiccated academic exchange between commentators" (Richardson, 1996, p. 35). The existing literature gives several ways for groups to act as change agents, triggering a process of change:

- by framing the range of political controversy surrounding an issue/diffusing new ideas and information, thereby teaching policy makers new ideas and information, which can change State interest and objectives
- by setting standards
- by controlling knowledge and information (Mazey & Richardson, 1996, p. 54), seeking to manipulate the rules, budgets, and personnel of governmental institutions to achieve common goals over time (Sabatier, 1993, p. 5).

Neofunctionalism for example sees groups as part of the process of political spill over (George, 1996, p. 13).

The degree of influence of groups to act as change agents is dependent on their access to decision-makers and epistemic communities (Suzuki, 2003, p. 19). Their influence as change agents is particularly high under conditions of uncertainty,[19] which give rise to policy-makers demand for information, provided by groups such as epistemic communities. Policy makers turn to new and different channels of advice when faced with uncertainty (Richardson, 1996, p. 38). According to Haas (1992, p. 12) many of the conditions facilitating a focus on power are absent in the face of uncertainty. Political leaders find it hard to identify political allies (Richardson, 1996, p. 38). Among the factors contributing to the uncertainties faced by decision makers, as mentioned by Haas (1992), "are the increasingly complex and technical nature of the ever-widening ranges of issues" (p. 12) under consideration. The potential of groups to influence the policy process is dependent on their knowledge, authoritativeness, and legitimacy (Richardson, 1996, p. 40). In this context, being of

[18]For a discussion on the use of the model of advocacy coalitions cf. Dudley and Richardson (1999).

[19]Uncertainty refers to the extent to which actors are not fully informed about others' behavior, the state of the world, and/or others' preferences. cf. Haas (1992, p. 3).

technical nature, space is an area of high uncertainty, where governments need great advice from experts. Experts can transform knowledge into power, pursuing their own personal and organisational goals (Suzuki, 2003, pp. 16–7).

According to Sabatier (1998, p. 119) groups can also form grand coalitions. He identifies three conditions for a successful consensus process:

- a stalemate wherein all coalitions view a continuation of the status quo as unacceptable;
- the negotiations are conducted in private and last a relatively long time, e.g. more than six months;
- there is a facilitator (policy broker) respected by all parties and viewed as relatively neutral (Sabatier, 1998, p. 119).

According to Marta and Stephenson (2016, p. 106) the Commission has long dealt with private firms operating as part of a "space lobby". These include large transnational aerospace and defence companies such as Airbus, Thales Alenia Space, Finmeccanica, GMV and Indra as well as companies benefitting from EU contracts such as OHB. Moreover, hundreds of small and medium-sized enterprises with unique and extremely technical and refined know-how are involved as suppliers. They are organised within the Aerospace and Defence Industries Association of Europe (ASD) with a specific group dedicated to space called Eurospace. At the same time the European Satellite Operators' Association gathers industries involved in the delivery of telecommunication services. Trade associations coordinate among their European members for a common lobbying effort in the institutions (Ibid.).

2.4.4.4 Institutions and Institutional Interrelation

The fourth so-called change agent identified by Madariaga (2011) is institutions. Given the differentiation between endogenous and exogenous sources of institutional change, there are two ways for institutions to act as change agents (cf. 2.3.3 sources of change). First, institutions can be the affected actor changing due to endogenous sources such as self-undermining processes reaching a critical level (Greif & Laitin, 1994, p. 634), inherent ambiguities and "gaps" that exist by design or emerged over time" (Streeck & Thelen, 2005, p. 19), "internal contradictions" resulting from institutional design which lead "to serious failures mounting pressure for major alterations" (Young, 2010, p. 174). Secondly, institutions can impose change on another institution, acting as the designing or implementing actor (exogenous source).

According to theoretic accounts such as Neofunctionalism, Supranational Governance and Rational Choice Institutionalism supranational institutions are not neutral but exert significant independent influence on institutional and policy outcomes (Hix, 2005, p. 16), act as driving forces (Pierson, 1998, p. 29) and play an autonomous role (Andersen et al., 2001, p. 28). Supranational Governance (Marks et al., 1998; Pierson, 1996; Sandholtz & Stone Sweet, 1998; Jachtenfuchs, 2001; Kohler-Koch, 2000) explains institutional change as an outcome of the agenda-setting by supranational institutions rather than by Member States (Hix, 2005, p. 17; cf. Branch & Ohrgaard, 2011).

The existing literature gives several ways for institutions to act as change agents, triggering a process of change:

- by using their (formal or informal) powers to foster their own interest as an institution (Dehousse & Magnette, 2006, p. 30),
- by shaping the ways in which preferences, priorities and understanding of what is possible are developed (Vanhoonacker, 2005, p. 69),
- by controlling knowledge and information (Mazey & Richardson, 1996, p. 54),
- by taking a life of their own (Pollack, 1996, p. 429; March & Olsen, 2006, p. 12),
- by enabling, empowering and constituting actors who seek to change institutions (Campbell, 2004, p. 72),
- by competence maximisation (one institution can change another one by seeking to increase its own competences) (Pollack, 1997b, p. 3),
- by becoming self-undermining (Streeck & Thelen, 2005, p. 19).

Institutions can also be coupled (Madariaga, 2011, p. 25), meaning they complement each other when the presence or absence of one institution affects the efficiency of the other (Hall & Soskice, 2001). As a result, a shift in one institution can result, due to this complementarity, in a shift in the other (Ibid.). According to Tudyka (2005, p. 239) States can also use institutions as change agents. This process is referred to as "hegemonialisation". According to the analyses of Rational-Choice-Institutionalists, institutions act within the constraints of Member State's preferences (Pollack, 1997a, p. 129).

At the same time, in line with Historical Institutionalism institutions also constrain the ability of actors to change institutions by limiting the range of institutional design options and choices available to decision makers (Campbell, 2004, pp. 25–6). Past institutional choices are understood to shape trajectories of future change, or limit the possibility of change (Andersen et al., 2001, p. 28).

Considering one of the research questions under discussion regarding the impact of the European integration process on existing institutional actors, in line with Doleys (2000) the question arises in what situations governments can be expected to dominate policy-making, and under what conditions supranational actors can be expected to affect change independently of and even in the face of opposition from governments (Doleys, 2000, p. 546).

According to the literature under discussion, the scope for institutions to play an autonomous role is dependent on

- their ability to use informational and ideational asymmetries to overcome domestic and transnational coordination problems, serving as impartial mediators (Moravcsik, 1999),
- their degree of control of information,
- their capacity to strategically shape their proposals in ways that will secure approval, even if the subject-matter conflicts with the preferences of many Member States, incl. techniques such as packaging (i.e. manipulation of the cost-benefit

calculations by linking measures), framing, co-optive justification and camouflage (Tallberg, 2000, pp. 848–9),

- their leadership resources (levels of technical and legal knowledge of issue under discussion, i.e. content expertise; analytical skills; knowledge of the state of play of negotiations), the degree of organisation, i.e. the structure and competence of the institutions,
- the negotiating context,
- their choice of strategy in the negotiations (Beach, 2005, p. 1),
- a particular set of preferences, i.e. on the preference profile, of the individuals involved (Hammond, 2005, p. 128),
- the delegated authority of the institution, their longevity, leadership and legitimacy (Hix, 2005, p. 16), "institutional standing" and "institutional embeddedness",[20]
- institutional rules established for applying sanctions, overruling legislation and changing agents' mandates,
- the role of incomplete information and uncertainty in principal–agent relationships,
- the difficulty of reforming institutions (Andersen et al., 2001, p. 29).

Member States' monitoring and sanctions further influence the degree for supranational actors to be able to advance their own agenda (Tallberg, 2000, p. 850).

Pollack (1997a) concludes, "the precise causal role of the European Commission in the process of European policy-making and European integration remains theoretically contested and empirically unmapped" (Pollack, 1997c, p. 111). While the Commission's precise role has remained in dispute, the very notion of it acting as an agent has become "a semantic bone of contention" (Pollack, 1997c, p. 111) between Neofunctionalism and Intergovernmentalism. In order to advance the debate on the question of when supranational actors are able to advance their own agenda, this dissertation will focus on the EU's Commission among the EU institutions[21]. With its right of initiative[22] it may play a role by proposing a programme of institutional change (Pollack, 1996, p. 439; Dehousse & Magnette, 2006, p. 23), or shape an agreement by acting as a mediator between national preferences (Dehousse & Magnette, 2006, p. 31) and is thus often taken (e.g. Egeberg, 2004; Hooghe, 2001) as a prime example for institutional influence. Richardson (1996) for example refers to it as an "enormously powerful" (p. 36) institutional actor in the EU policy process, while Kaunert (2010) discussed the Commission's role as a supranational policy entrepreneur (Kaunert, 2010). According to Kingdon (1984) a policy entrepreneur, standing at a window of opportunity, invests all its resources such as time, reputation and money in order "to propose, lobby for and sell a policy proposal" (p. 188). According to Kassim and Dimitrakopolous (2007, p. 1251) the dominant image of

---

[20]Institutional embeddedness captures the relationship whether complementary or substitutive that a given international institution has with other institutions that appear to serve to the same purpose (Conner, 2010, p. ix).

[21]For more information on the EU institutions cf. Wallace (2000a), Dinan (1999).

[22]For more information on the role of the Commission refer to Ludlow (1991). For more information on the structure of the Commission and the policy formation, cf. Donnelly (1993).

the European Commission is that of an imperialistic bureaucracy, driven by an "institutional commitment to deeper integration" (Dinan, 1997, p. 261).

These claims face strong opposition by arguments such as "notions about the Commission, acting forceful as a unitary actor, are misleading" (Pollack, 1997c, p. 4; Hooghe, 2001, p. 193). The latter draw on the EU's Commission institutional setup, claiming the Commission's administrative and social political preferences vary (Hooghe, 2001, p. 193) due to the competing and diverging DGs (Pollack, 1997c, p. 4). The Commission is rather said to be culturally diverse and politically divided (Stephenson, 2012, p. 87; Kassim, Connolly, Dehousse, Rozenberg, & Bendjaballah, 2016, p. 5). Indeed, according to Kassim and Dimitrakopolous (2007, p. 1251), studies of Commission preferences at successive round of treaty reform reveal a more complex picture. In line with framing theory, frames may compete within an institution, meaning that multiple often conflicting, frames circulate in the policy-making environment, each informed by one or more sets of experts and interest groups. An analysis of frames requires an understanding of the institutional architecture of the "framers" in order to map institutional discourses and how they emerge (Stephenson, 2012, p. 33; p. 87). According to Tallberg (2000, p. 848) the Commission's chances of advancing its own interests are dependent on its capacity to strategically shape its proposals in ways that secure approval. Also, internal divisions limit the degree of influence (Pollack, 1997c, p. 51; Hooghe, 2001, p. 193). At the same time the Commission might also purposefully exploit these internal divisions "like running several horses in the race" (Stephenson, 2012, p. 93), trying to see which one can secure sufficient political and financial spots to result in a decision and implementation (Ibid.).

In terms of resources at its disposal it can be differentiated between legal, political and financial ones. Many DGs are involved in space policy. In terms of human resources, the Commission has limited expertise, being composed mainly of policy-makers who are engaged in elements of policy formulation, lacking technical expertise. Thus the Commission relies on external scientific and technical expertise, industrial lobbies as well as ESA. It seeks to shape and steer the agenda, defining objectives, identifying policy priority and providing impetus and momentum.[23] Hartlapp et al. (2013) have analysed the bureaucratic politics inside the European Commission and conclude "that some Commission actors hold advantages over others when internal conflict on regulatory options exist." (p. 431) Accordingly especially the leading DG can establish tight relations with external stakeholder and is able to build a stock of relevant information during the first stages of policy formulation. Additionally, it is this DG, who decides, which other DGs to get involved. The bureaucratic impact of the lead DG strongly impacts on the legislative agenda proposed by the European Commission. The Secretariat General of the Commission in turn acts as a "gate keeper" (Borràs, 2007, p. 5) in allowing "the extent to which particular bureaucratic interests can bias the legislative agenda proposed by the Commission." (Hartlapp et al. 2013, p. 431). Considering the Common Agricultural Policy as an example,

[23] For a more detailed discussion on the role of the European Commission in Space policy cf. Marta and Stephenson (2016, p. 101).

Blom-Hansen (2008, p. 212, 223) concludes that when Member States have diverging preferences, the Commission's proposal represented a solution on which intergovernmental bargaining converged.

2.4.5 *Forms of Institutional Change*

In addition to the previously identified facets of institutional change, it is possible to distinguish different forms of institutional change, which are also often referred to as outcomes of institutional crisis. The existing literature refers to the following outcomes of institutional change:

- latent institutions become salient (Suzuki, 2003, p. 33) and are dissolved;
- institutions are put in the service of different ends (Ibid.);
- adjustment or learning: an existing institution is adjusted (Ibid.)
  - marginal adaptation to environmental changes without changing core principles;
  - gradual adaptation through partial renewal of institutional arrangements and limited redirection of core principles (Ebbinghaus, 2005, p. 17);
- bricolage: institutional change involving the "recombination of old institutional elements" (Campbell, 2004, p. xviii)
  - substantive bricolage, involving the recombination of already existing institutional principles and practices, and
  - symbolic bricolage, involving the recombination of symbolic principles and practises (Ibid., pp. 69–70);
- translation
  - introduction of new institutional elements (Streeck & Thelen, 2005, p. 21)
  - displacement: a new institution is founded (March & Olsen, 1989, p. 169; Suzuki, 2003, p. 33; Ebbinghaus, 2005, p. 17).

According to Olsen (1997) institutional change can take place through adjustment or learning, without a decision for change as "institutions are often transformed through mundane processes of learning and incremental adaptation" (p. 209). He links the concept of learning to the concept of institutional design, saying that "imperfect learning routines are a precondition for radical design". The more efficient ordinary processes of learning and adaptation are the less likely is accumulation of tensions and conflicts and radical change." (pp. 209–10). Thelen (1999) refers to the "mechanisms of reproduction", arguing that all institutions are continually renegotiated and reinterpreted in order to adapt to continuous changing circumstances, and remain socially embedded (p. 387). Florensa (2004) refers to "learning processes, by the acquisition of new skills, information and knowledge by the actors involved lead to readjustments in the institutional structure" (p. 9) as the third stage of the constant institutional development.

According to several authors (cf. Campbell, 2004, p. 72; Jupille, 2004, p. 1; Lanzara, 1998, p. 12) actors aiming to change an institution are constrained by the availability of institutional alternatives. Accordingly, institutions do not emerge in a vacuum but always "challenge, borrow from, and, to varying degrees, display prior institutions" (Scott, 2008, p. 94). March and Olsen (2006, p. 11) thus refer to a competitive selection of elements of institutions. Campbell (2004) refers to this phenomenon as "bricolage" and differentiates between substantive bricolage, involving the recombination of already existing institutional principles and practices, and symbolic bricolage, involving the recombination of symbolic principles and practises (pp. 69–70). Bricolage allows economising sunk costs, by converting old components and structures[24] to new functions and uses (Lanzara, 1998, p. 29). By providing the tool kit or repertoire for actors to modify institutions, institutions also constrain the process of change as the number of elements is limited. According to Lanzara (1998, p. 11), bricolage is often chosen as the form of change, when perceived uncertainty and instability become intolerable for actors, as they give up exploring and fall back to reinstating familiar arrangements. According to Campbell (2004, p. 74), bricolage is also often chosen by individuals acting as change agents, as creative and innovative people (bricoleurs or entrepreneurs) tend to recombine institutional elements in innovative ways. Additionally, the discretion and choice of institutional designers are constrained by institutional environments, i.e. "collective moral and causal beliefs providing organisational "templates" and "scripts" (Ibid., pp. 205–6).

Another form of institutional change is "translation", involving the introduction of new institutional elements (Ibid., p. 28). A radical form of translation is displacement. According to Streeck and Thelen (2005), displacement happens "as new models emerge and diffuse, which call into question existing, previously taken-for-granted organizational forms and practices" (p. 19). Change through displacement can occur endogenously through the rediscovery or activation of previously suppressed or suspended possibilities (Ibid., p. 21).

2.5 Sub-conclusion

As has been shown the existing literature seems to lack a structured approach to institutional change. Several authors have called for an analysis of the factors most important in describing institutional change (Keohane & Nye, 2001, p. 50) and more research on the development of institutions once they have been put into place (Héritier, 2007, p. 1). In line with this Shepsle (1989) underlined the need to understand

[24]Bureaucrats hold institution-specific knowledge and practical experience acquired through a long and costly learning process, which makes the establishment of new institutions difficult, costly and dilemmatic: "if existing competencies and human resources are kept out, institution building suffers from lack of experienced personnel, but if it is called in, they tend to implant the old rules, practices and habits, sabotaging the innovative effort and reinstating many of the features of the previous institutional regime". Cf. Lanzara (1998, p. 13).

the selection, survival, adaptation, and evolution of institutional practices (Shepsle, 1989a, p. 133) and Campbell (2004) suggested to identify different types of change first (Campbell, 2004, p. 31). Sigalas (2016), who analysed the role of the European Parliament in framing space policy, even called for "a fully-fledged theory of change" (Sigalas, 2016, p. 77).

To theoretically contribute to our understanding of institutional change, to advance the debate about the impact of the European integration process on existing institutional actors and to make a theoretical contribution to our understanding of when supranational actors are able to advance their own agenda, this chapter has reviewed the state of the art of literature dealing with institutional change and the European integration process. It gave an overview of the applicable tenants of European Integration Theories (i.e. Neofunctionalism, Intergovernmentalism and Liberal Intergovernmentalism, Supranational Governance), Agency Theories (cf. Barbieri & Ongaro, 2008), Theories of International Relations and New Institutionalism. They have been scrutinised with the objective to identify the factors most important in describing institutional change and to offer a more structured approach to the analysis of institutional change as to be able to compare different cases.

To structure observations on institutional change, the existing theoretical frameworks were considered as complementary, answering complementary aspects of the same puzzle. It is proposed to analyse cases of institutional change in terms of the above-mentioned stages of institutional development and facets of institutional change. In line with Keohane and Nye (2001, p. 50) these are suggested as the most important factors describing institutional change. By applying them to different cases, they will become comparable.

To make a theoretical contribution to our understanding of institutional change, the question arises whether the proposed five facets of institutional change prove to be suited to structure the debate. It is expected that the differentiation regarding the degree and pace of institutional change might be difficult to apply to the case study or be still too broad. Moreover, on the basis of the literature outlined above, conditions under which endogenous or exogenous sources preside could not be identified. The case study might give further insights. The analysis of the existing theoretical frameworks on institutional change has highlighted the existence of four so-called change agents, activating sources of institutional change. The following conditions working in favour or against a change agent to advance its own agenda could be identified.

In line with many Historical Institutionalists (i.e. North, 1990; March & Olsen, 1989, p. 169, Pollack, 1996; Campbell, 2004), it is expected to find that most of the six phases of institutional development of space policy in Europe have undergone incremental change. Probably both endogenous (i.e. Greif & Laitin, 1994; Streeck & Thelen, 2005; Young, 2010) and exogenous (Pollack, 1996, p. 441; Krasner, 1988, p. 67; Scott, 2008; Steinmo et al., 1992; Suzuki, 2003) sources of institutional change (Florensa, 2004, p. 6) can be identified in the case study. Furthermore, the case might give insights which change agents preside under which conditions. According to the literature outlined above, the following matrix explains the conditions working in favour of and against the influence of a particular change agent (Fig. 2.1).

It is expected to find out that States acted as change agents activating institutional change in all six phases under discussion. Especially as space policy is a matter of national sovereignty, which again is a symbol of scientific and technical capability, an index of political power and a component of military strategy, and thus governed by States (Wörner, 2013). In line with North (1990) and Fligstein (2010) who have considered so-called "institutional entrepreneurs" as the main source for institutional change it will be interesting to see in how far and when individuals can act as change agents. It is also likely, in line with Haas (1992, p. 12), that this study finds proofs for groups acting as change agents, as space policy is a policy areas which is highly technical, giving rise to policy-makers' demand for information, which can be provided by expert groups such as epistemic communities. Given that both ESA and the EU operate in the same geographic realm, it can be expected that institutions of the EU, i.e. the Commission as policy entrepreneur, will be found to have acted as a change agent.

While theoretically it seems to be relatively easy to distinguish between the four different results of institutional change, i.e. institutional adjustment, bricolage, translation, displacement, it will be shown that in practice it is relatively more complicated to categorise institutional change in these four options.

Conditions for Actors to be able to advance their own agenda	Change Agents			
	States	Supranational Actors / Institutions	Groups	Individuals
favourable	• bargaining power • issues of national sovereignty	• authority • legitimacy • "institutional standing" • "institutional embeddedness" • control of information and knowledge • ability to abuse information asymmetries • ability to frame / shape a proposal • combined with powerful allies such as States	• uncertainty/ need for knowledge and information • ability to frame an issue • authoritativenes s / legitimacy • ability to access decision-makers	• social capital • ability to access decision makers
unfavourable	• uncertainty	• fragmentation • uncertainty	• not identified	• not identified

Fig. 2.1 Favourable and unfavourable factors for each change agent to be able to advance its own agenda

References

Abbott, K. W., & Snidal, D. (2010). Why states act through formal international organizations. In J. L. Goldstein & R. H. Steinberg (Eds.), *International Institutions: Vol. II. Consequences: When, where and why international institutions are effective* (pp. 239–71). London: SAGE.

Ackrill, R., & Kay, A. (2011). Multiple streams in EU policy-making: The case of the 2005 sugar reform. *Journal of European Public Policy, 18*(1), 72–89.

Aliberti, M., & Krasner, S. D. (2016). Governance in space. In C. Al-Ekabi, B. Baranes, P. Hulsroj, & A. Lahcen (Eds.), *Yearbook on space policy 2014* (pp. 143–166). Vienna: Springer.

Aminzade, R. (1992). Historical sociology and time. *Sociological Methods & Research, 20,* 456–480.

Andersen, S. S., Eliassen, K. A., & Sitter, N. (2001). Formal processes: EU institutions and actors. In S. S. Andersen & K. A. Eliassen (Eds.), *Making policy in Europe* (pp. 20–43). London: SAGE Publisher.

Aspinwall, M. D., & Schneider, G. (2000). Same menu, separate tables: The institutionalist turn in political science and the study of European integration. *European Journal of Political Research, 38,* 1–36.

Barbieri, D., & Ongaro, E. (2008). EU agencies: What is common and what is distinctive compared with national-level public agencies. *International Review of Administrative Sciences, 74*(3), 395–420.

Barnett, M., & Finnemore, M. (1999). The politics, power, and pathologies of international organizations. *International Organization, 53*(4), 699–732.

Baumgartner, F., & Jones, B. D. (1993). *Agendas and instability in American Politics*. Chicago: University of Chicago Press.

Beach, D. (2005). *The dynamics of European integration—why and when EU institutions matter*. Houndmills: Palgrave Macmillan.

Bendor, J., Glazer, A., & Hammond, T. H. (2000). *Theories of delegation in political science*. Retrieved from https://www.gsb.stanford.edu/faculty-research/working-papers/theories-delegation-political-science. Accessed January 18, 2012.

Blom-Hansen, J. (2008). The origins of the EU comitology system: A case of informal agenda-setting by the commission. *Journal of European Public Policy, 152,* 208–226.

Boin, A. (2008). Mapping trends in the study of political institutions. *International Studies Review, 10,* 87–92.

Borrás, S. (2007). The european commission as network broker. *European Integration Online Papers, 11*(1).

Börzel, T. A. (1998). Organizing Babylon—On the different conceptions of policy networks. *Public Administration, 76,* 253–273.

Brachet, G. (2016). The optimum role of governments in space. In C. Al-Ekabi, B. Baranes, P. Hulsroj, & A. Lahcen (Eds.), *Yearbook on space policy 2014* (pp. 211–215). Vienna: Springer.

Branch, A. P., & Ohrgaard, J. C. (2011). Trapped in the supranational-intergovernmental dichotomy: A response to Stone Sweet and Sandholtz. *Journal of European Public Policy, 6*(1), 123–143.

Bulmer, S. (1994). The governance of the european union: A new institutionalist approach. *Journal of Public Policy, 13*(4), 352–380.

Bulmer, S., & Padgett, S. (2004). Policy transfer in the European union: An institutionalist perspective. *British Journal of Political Sciences, 35,* 103–126.

Buzan, B., Jones, C., & Little, R. (1993). *The logic of anarchy—Neorealism to structural realism*. New York: Columbia University Press.

Campbell, J. (2004). *Institutional change and globalization*. Princeton: Princeton University Press.

Caporaso, J. (1998). Regional integration theory: Understanding our past and anticipating our future. *Journal of European Public Policy, 5*(1), 1–16.

Christensen, J. G., & Nielsen, V. L. (2010). Administrative capacity, structural choice and the creation of EU-agencies. *Journal of European Public Policy, 17*(2), 176–204.

Christiansen, T., & Vanhoonacker, S. (2008). At a critical juncture? Change and continuity in the Institutional development of the council secretariat. *West European Politics, 31*(4), 751–770.
Cobb, R. W., & Elder, C. D. (1971). The politics of agenda-building: An alternative perspective for modern democratic theory. *The Journal of Politics, 33*(4), 892–915.
Conner, A. (2010). Introduction: International institutions and effectiveness—Achieving cooperation. In J. L. Goldstein & R. H. Steinberg (Eds.), *International Institutions: Vol. II. Consequences: When, where and why international institutions are effective* (pp. iiix–x). London: SAGE.
Copeland, P., & James, S. (2014). Explaining the relaunch of the EU's economic reform agenda. *Journal of European Public Policy, 21*(1), 1–19.
Cram, L. (1993). Calling the tune without paying the piper: Social policy regulation: The role of the commission in european union social policy. *Policy and Politics, 21*(2), 135–146.
Daviter, F. (2007). Policy framing in the European union. *Journal of European Public Policy, 14*(4), 654–666.
Daviter, F. (2012). Framing biotechnology policy in the European union. *ARENA Working Paper*. Retrieved from http://www.sv.uio.no/arena/english/research/publications/arena-working-papers/2012/wp5-12.pdf. Accessed March 25, 2017.
Dehousse, R., & Magnette, P. (2006). Institutional change in the EU. In J. Peterson & M. Shackleton (Eds.), *The institutions of the European union* (pp. 17–34). Oxford: Oxford University Press.
DiMaggio, P. (1988). Interest and agency in institutional theory. In L. G. Zucker (Ed.), *Institutional patterns and organizations: Culture and environment* (pp. 3–23). Cambridge: Ballinger Pub.
Dinan, D. (1997). The commission and the intergovernmental conferences. In N. Nugent (Ed.), *At the heart of the union* (pp. 254–264). Basingstoke: Macmillan.
Dinan, D. (1999). *Ever closer union—An introduction to European integration*. Houndmills: Palgrave.
Doleys, T. (2000). Member states and the European commission: Theoretical insights from the new economics of organization. *Journal of European Public Policy, 7*(4), 532–553.
Donnelly, M. (1993). The structure of the European commission and the policy formation process. In S. Mazey & J. Richardson (Eds.), *Lobbying in the European union* (pp. 74–81). Oxford: Oxford University Press.
Dudley, G., & Richardson, J. (1999). Competing advocacy coalitions and the process of "frame reflection": A longitudinal analysis of EU steel policy. *Journal of European Public Policy, 6*(2), 225–248.
Duke, S., & Vanhoonacker, S. (2006). Administrative governance in the CFSP: Development and practice. *European Foreign Affairs Review, 11,* 163–182.
Ebbinghaus, B. (2005). Can path dependence explain institutional change—Two approaches applied to welfare state reform. *MPlfG Discussion Paper 05/2*. Retrieved from http://www.mpifg.de/pu/mpifg_dp/dp05-2.pdf. Accessed March 17, 2017.
Edelmann, R., Milde, H., & Weimerskirch, P. (1998). Agency-Beziehungen und Kontrakt-Design: Problem, Lösung, Beispiel. *Kredit und Kapital, 1,* 1–27.
Egeberg, M. (1994). Bridging the gap between theory and practice: The case of administrative policy. *Governance: An International Journal of Policy, Administration, and Institutions, 7.1,* 83–98.
Egeberg, M. (2004). An organisational approach to European integration: Outline of a complementary perspective. *European Journal of Political Research, 43,* 199–219.
Egeberg, M., & Trondal, J. (2010). *EU-level agencies: New executive centre formation or vehicles for national control?* Paper presented at the ECPR Fifth Pan-European Conference, Porto 24–26 June 2010. Retrieved from http://www.jhubc.it/ecpr-porto/virtualpaperroom/010.pdf. Accessed March 17, 2017.
Fligstein, N. (1997a). Social skill and institutional theory. *American Behavioral Scientist, 40*(4), 397–405.
Fligstein, N. (1997b). *Social skill and the theory of fields*. Retrieved from http://www.irle.berkeley.edu/culture/papers/Fligstein01_01.pdf. Accessed March 13, 2017.

Fligstein, N. (2010). Institutional entrepreneurs and cultural frames—The case of the European union's single market program. *European Societies, 3*(3), 261–287.
Florensa, M. C. (2004). *Institutional stability and change. A logic sequence for studying institutional dynamics*. Paper presented at the Tenth Biennial Conference of the International Association for the Study of Common Property (IASCP), Oaxaca, Mexico. Retrieved from https://dlc.dlib.indiana.edu/dlc/bitstream/handle/10535/1823/CostejaFlorensa_Institutional_040525_Paper220.pdf?sequence=1. Accessed March 13, 2017.
George, S. (1996). The European union—Approaches from international relations. In H. Kassim & A. Menon (Eds.), *The European union and national industrial policy* (pp. 11–25). London: Routledge.
Geradin, D., & Petit, N. (2004). The development of agencies at EU and national levels: Conceptual analysis and proposals for reform. *Jean Monnet Working Paper 1*. Retrieved from http://www.jeanmonnetprogram.org/archive/papers/04/040101.pdf. Accessed March 18, 2017.
Giannopapa, C., Adriaensen, M., & Sagath, D. (2016). The Member states of the european space agency—National governance structures, priorities and motivations for engaging in space. In T. Hörber & P. Stephenson (Eds.), *European space policy* (pp. 173–190). London: Routledge.
Goldstein, J. L., & Steinberg, R. H. (2010). Introduction. In J. L. Goldstein & R. H. Steinberg (Eds.), *International Institutions. Vol. I: Causes* (pp. xvii–xxxi). London: SAGE Publisher.
Greif, A., & Laitin, D. D. (1994). A theory of endogenous institutional change. *American Political Science Review, 98*(4), 633–652.
Haas, P. M. (1992). Introduction: Epistemic communities and international policy coordination. *International Organization, 46*(1), 1–35.
Haas, P. M. (2010). Introduction: Epistemic communities and international policy coordination. In J. L. Goldstein & R. H. Steinberg (Eds.), *International institutions* (Vol. I, pp. 185–220). London: SAGE.
Hall, P. A. (1993). Policy paradigms, social learning, and the state: The case of economic policymaking in Britain. *Comparative Politics, 25*(3), 275–296.
Hall, P. A., & Soskice, D. (2001). *Varieties of capitalism: The institutional foundations of comparative advantage*. Oxford: Oxford University Press.
Hall, P. A., & Taylor, R. C. R. (1996a). Political science and the three new institutionalisms. *Political Studies, XLIV*, 936–57.
Hall, P. A., & Taylor, R. C. R. (1996b). *Political science and the three new institutionalism*. Retrieved from http://www.mpifg.de/pu/mpifg_dp/dp96-6.pdf. Accessed March 13, 2017.
Hammond, T. H. (2005). Formal theory and the institutions of governance. *Governance, 9*(2), 107–185.
Hartlapp, M., Metz, J., & Rauh, C. (2013). Linking agenda setting to coordination structures: Bureaucratic politics inside the European commission. *Journals of European Integration, 35*(4), 425–441.
Héritier, A. (2007). *Explaining institutional change in Europe*. New York: Oxford University Press.
Hix, S. (2005). *The political system of the European union*. Houndmills: Palgrave.
Hooghe, L. (2001). *The European commission and the integration of Europe*. Cambridge: Cambridge University Press.
Horak, M. (2007). *Governing the post-communist city: Institutions and democratic development in Prague*. Toronto: University of Toronto.
Hörber, T., & Stephenson, P. (Eds.). (2016). *European space policy*. London: Routledge.
Ikenberry, G. J. (1994). *History's heavy hand: Institutions and the politics of the state*. Paper prepared for a conference on "New Perspectives on Institutions", University of Maryland. Retrieved from https://scholar.princeton.edu/sites/default/files/HistorysHeavyHand_0.pdf. Accessed March 13, 2017.
Jachtenfuchs, M. (2001). The governance approach to European integration. *Journal of Common Market Studies, 39*(2), 245–264.
Jupille, J. (2004). *Procedural politics—Issues, influence, and institutional choice in the European union*. Cambridge: Cambridge University Press.

Kassim, H., & Dimitrakopoulos, D. G. (2007). The commission and the future of Europe. *Journal of European Public Policy, 14*(8), 1249–1270.

Kaunert, C. (2010). The area of freedom, security and justice in the Lisbon Treaty: Commission policy entrepreneurship? *European Security, 19*(2), 169–189.

Keohane, R., & Nye, J. (2001). *Power and interdependence*. New York: Longman.

Kingdon, J. W. (1984). *Agenda, alternatives and public policies*. Boston: Little Brown.

Kohler-Koch, B. (2000). Framing: The bottleneck of constructing legitimate institutions. *Journal of European Public Policy, 7*(4), 513–531.

Koremenos, B., Lipson, C., & Snidal, D. (2001). The rational design of international institutions. *International Organization, 55.4*(Autumn), 761–799.

Krasner, S. (1988). Sovereignty—An institutional perspective. *Comparative Political Studies, 21*(1), 66–94.

Krasner, S. (2010). Global communication and national power: Life on the Pareto frontier. In J. L. Goldstein & R. H. Steinberg (Eds.), *International Institutions* (Vol. II, pp. 1–28)., Consequences: When, where and why International Institutions are effective London: SAGE Publisher.

Lake, D. A. (2001). Beyond anarchy: The importance of security institutions. *International Security, 26*(1), 129–160.

Lanzara, G. F. (1998). Self-destructive processes in institution building and some modest countervailing mechanisms. *European Journal of Political Research, 33,* 1–39.

Levy, M. A., Young, O. R., & Zürn, M. (1995). The study of international regimes. *European Journal of International Relations, 1,* 267–330.

Ludlow, P. (1991). The European commission. In R. O. Keohane & S. Hoffmann (Eds.), *The new European community. Decision-making and Institutional Change* (pp. 85–132). Boulder: Westview Press.

Madariaga, A. (2011). *Patterns of institutional change and external competitiveness in Neoliberal and dependent political economies. The Cases of Chile and Estonia*. Budapest: Central European University.

Majone, G. (2001). Two logics of delegation: Agency and fiduciary relations in EU governance. *European Union Politics, 2,* 103–122.

March, J. G., & Olsen, J. P. (1989). *Rediscovering institutions: The organizational basis of politics*. New York: The Free Press.

March, J. G., & Olsen, J. P. (1998). The institutional dynamics of international political orders. *International Organization, 52*(4), 943–969.

March, J. G., & Olsen, J. P. (2006). Elaborating the 'New Institutionalism'. In R. A. W. Rhodes, S. A. Binder, & B. A. Rockman (Eds.), *The oxford handbook of political institutions* (pp. 3–20). Oxford: Oxford University Press.

Marta, L., & Stephenson, P. (2016). Role of the European commission in framing European space policy. In T. Hörber & P. Stephenson (Eds.), *European space policy* (pp. 98–113). London: Routledge.

Marks, G., Scharpf, F. W., Schmitter, P. C., & Streeck, W. (1998). Governance in the European Union. London: SAGE.

Mazey, S., & Richardson, J. (1996). EU policy-making. A garbage can or an anticipatory and consensual policy style? In Y. Mény, P. Muller, & J. Quermonne (Eds.), *Adjusting to Europe—The impact of the European Union on national institutions and Policies* (pp. 41–58). London: Routledge.

Mearsheimer, J. (1994). The false promise of International Institutions. *International Security, 19*(3), 5–49.

Meunier, S., & McNamara, K. R. (2007). *Making history: European integration and institutional change at fifty*. Oxford: Oxford University Press.

Moravcsik, A. (1993). Preferences and power in the European community: A liberal intergovernmentalist approach. *Journal of Common Market Studies, 31*(4), 473–522.

Moravcsik, A. (1998). *The choice for Europe: Social purpose and state power from Messina to Maastricht*. London: Cornell University Press.

Moravcsik, A. (1999). A new statecraft? Supranational entrepreneurs and international cooperation. *International Organization, 53*(2), 267–306.

North, D. C. (1990). *Institutions, institutional change and economic performance*. Cambridge: Cambridge University Press.

Olsen, J. P. (1997). Institutional design in democratic context. *The Journal of Political Philosophy, 5*(3), 203–229.

Peterson, J., & Bomberg, E. (1999). *Decision-making in the European union*. New York: St. Martin's Press Inc.

Peterson, J., & Shackleton, M. (2006b). The EU's institutions. An overview. In J. Peterson & M. Shackleton (Eds.), *The institutions of the European union* (pp. 1–16). Oxford: Oxford University Press.

Pierson, P. (1994). The path to european integration: A historical institutionalist perspective. Program for the Study of Germany and Europe, Working Paper No. 5.2. Retrieved from http://aei.pitt.edu/63633/1/PSGE_WP5_2.pdf. Accessed March 29, 2017.

Pierson, P. (1996). The New institutionalism and EC governance: The promise and limits of institutional analysis. *Governance: An International Journal of Policy, Administration, and Institutions, 9.4*, 429–458.

Pierson, P. (1998). The path to European integration: A historical institutionalist perspective. In W. Sandholtz & A. Stone Sweet (Eds.), *European integration and supranational governance* (pp. 27–58). Oxford: Oxford University Press.

Pierson, P. (2000). The limits of design: Explaining institutional origins and change. *Governance, 13*(4), 475–499.

Pollack, M. A. (1996). The new institutionalism and EC governance: The promise and limits of institutional analysis. *Governance: An International Journal of Policy, Administration, and Institutions, 9.4*, 429–458.

Pollack, M. A. (1997a). Delegation, agency, and agenda setting in the European community. *International Organization, 51*(1), 99–134.

Pollack, M. A. (1997b). *The engines of integration? Supranational autonomy and influence in the European Union*. Retrieved from http://aei.pitt.edu/2706/1/002752_1.PDF. Accessed March 17, 2017.

Pollack, M. A. (1997c). *The engines of integration? Supranational autonomy and influence in the European Union*. Retrieved from http://aei.pitt.edu/2706/1/002752_1.PDF. Accessed March 17, 2017.

Princen, S. (2007). Agenda-setting in the European union: A theoretical exploration and agenda for research. *Journal of European Public Policy, 14*(1), 21–38.

Princen, S. (2011). Agenda-setting strategies in eu policy processes. *Journal of European Public Policy, 18*(7), 927–943.

Puchala, D. J. (1999). Institutionalism, intergovernmentalism and European integration: A review article. *Journal of Common Market Studies, 37*(2), 317–331.

Putnam, R. D. (1988). Diplomacy and domestic politics: the logic of two-level games. *International Organization, 42*(3), 427–460.

Rein, M., & Schön, D. (1991). Frame-reflective policy discourse. In P. Wagner (Ed.), *Social sciences and modern states: National Experiences and theoretical crossroads* (pp. 261–289). Cambridge: Cambridge University Press.

Rhodes, R. A. W., Binder, S. A., & Rockman, B. A. (Eds.). (2006). *The oxford handbook of political institutions*. Oxford: Oxford University Press.

Richardson, J. (1996). Actor-based models of national and EU policy making. In H. Kassim & A. Menon (Eds.), *The European union and national industrial policy* (pp. 26–51). London: Routledge.

Rittberger, B., & Wonka, A. (2011). Introduction: Agency governance in the European Union. *Journal of European Public Policy, 18*(6), 780–789.

Rittberger, V., & Zangl, B. (2006). *International organization—Polity, politics and policies*. New York: Palgrave Macmillan.

Roland, G. (2004). Understanding institutional change: Fast-moving and slow-moving institutions. *Studies in Comparative International Development, 38*(4), 109–131.

Rosamond, B. (2000). *Theories of European integration*. Houndmills: Palgrave.

Sabatier, P. A. (1993). *Theoretical lenses on public policy*. Boulder: Westview Press.

Sabatier, P. A. (1998). The advocacy coalition framework: Revisions and relevance for Europe. *Journal of European Public Policy, 5*(1), 98–130.

Sabatier, P. A., & Jenkins-Smith, H. S. (1996). *The advocacy coalition framework: An assessment*. Paper presented to the Annual Meeting of the American Political Science Association.

Scharpf, F. W. (1997). *Games real actors play. Actor-centered institutionalism in policy research*. Oxford: Westview Press.

Scott, R. W. (2008). *Institutions and organizations—Ideas and interests*. London: SAGE Publications.

Shepsle, K. A. (1989). Institutional equilibrium and equilibrium institutions. *Journal of Theoretical Politics, 1*(2), 131–147.

Sigalas, E. (2016). Europe in space—The European parliament's justification arsenal. In T. Hörber & P. Stephenson (Eds.), *European space policy* (pp. 66–81). London: Routledge.

Sjöstrand, S. (1993). *Institutional change—Theory and empirical findings*. New York: M. E. Sharpe.

Snidal, D. (2010). Coordination versus prisoners' dilemma: Implications for international cooperation and regimes. In J. L. Goldstein & R. H. Steinberg (Eds.), *International Institutions: Vol. I. Causes* (pp. 2–31). London: SAGE Publisher.

Steinmo, S., Thelen, K., & Longstreth, F. (1992). *Structuring politics—Historical institutionalism in comparative analysis*. Cambridge: Cambridge University Press.

Stephenson, P. (2012). Talking space: The European commission's changing frames in defining Galileo. *Space Policy, 28*, 86–93.

Stephenson, P. (2016). Framing as a tool for analysing European space policy. In T. Hörber & P. Stephenson (Eds.), *European space policy* (pp. 30–49). London: Routledge.

Streeck, W., & Thelen, K. (2005). *Beyond continuity. Institutional change in advanced political economies*. Oxford: Oxford University Press.

Suzuki, K. (2003). *Policy logics and institutions of European space collaboration*. Burlington: Ashgate.

Sandholtz, W. & Stone Sweet, A. (1998). *European Integration and Supranational Governance*. Oxford: OUP.

Tallberg, J. (2000). The anatomy of autonomy: An institutional account of variation in supranational influence. *Journal of Common Market Studies, 38*(5), 843–864.

Tallberg, J. (2003). *European governance and supranational institutions—Making states comply*. London: Routledge.

Thelen, K. (1999). Historical institutionalism in comparative politics. *Annual Review of Political Sciences, 2*, 369–404.

Trondal, J., & Jeppesen, L. (2008). Images of agency governance in the European Union. *West European Politics, 31*(3), 417–441.

Tudyka, K. P. (2005). How the OSCE is hegemonalizing. In D. S. Lutz & K. P. Tudyka (Eds.), *Perspektiven und Defizite der OSZE* (pp. 239–250). Baden-Baden: Nomos Verlagsgesellschaft.

Vanhoonacker, S. (2005). The institutional framework. In C. Hill & M. Smith (Eds.), *International relations and the European union* (pp. 67–90). Oxford: Oxford University Press.

Wallace, H. (2000a). The policy process. In H. Wallace & W. Wallace (Eds.), *Policy-making in the European union* (pp. 39–64). Oxford: Oxford University Press.

Wallace, H. (2000b). The institutional setting. In H. Wallace & W. Wallace (Eds.), *Policy-making in the European union* (pp. 3–38). Oxford: Oxford University Press.

Waltz, K. N. (1979). *Theory of international politics*. New York: McGraw-Hill Publishing Company.

Warleigh, A. (2001). *Understanding European union institutions*. London: Routledge.

Wendt, A. (2001). Driving with the rearview mirror: On the rational science of institutional design. *International Organization, 55*(4), 1019–1049.

Wessel, W. (1997). An ever closer fusion? A dynamic macropolitical view on intergration processes. *Journal of Common Market Studies, 35*(2), 267–299.
Williamson, O. (1985). *The Economic institutions of capitalism*. New York: The Free Press.
Wonka, A., & Rittberger, B. (2010). Credibility, complexity and uncertainty: explaining the institutional independence of 29 EU agencies. *West European Politics, 33*(4), 730–752.
Wörner, J. (2013). *Von der kalten Kartoffel zum toten Pferd .... DLR Blogs – Jan Woerner's Blog*. Retrieved from http://www.dlr.de/blogs/de/desktopdefault.aspx/tabid-5896/9578_read-632/. Accessed March 18, 2017.
Yataganas, X. A. (2001). Delegation of regulatory authority in the European Union—The relevance of the American model of independent agencies. *Jean Monnet Working Paper 3*. Retrieved from http://www.jeanmonnetprogram.org/archive/papers/01/010301.html. Accessed March 18, 2017.
Yi-Chong, X., & Weller, P. (2008). To be, but not to be seen: Exploring the impact of international civil servants. *Public Administration, 86*(1), 35–51.
Young, O. R. (2010). Regime dynamics: The rise and fall of international regimes. In J. L. Goldstein & R. H. Steinberg (Eds.), *International Institutions: Vol. II. Consequences: When, where and why international institutions are effective* (pp. 163–83). London: SAGE Publisher.

Chapter 3
Empirical Study: The Development of Space Policy in Europe

Although the process of European economic integration has enjoyed success since the 1950s, cooperation in the field of space policy has lagged behind and was late to develop. This is striking, given that it perfectly fitted Monnet's definition of an ideal area for the advancement of European integration, as it was too big for individual nation States but also a new field of politics comparable with nuclear research under EURATOM (Hörber, 2016a, p. 53). In Europe space activities were originally conducted by individual States (Aliberti & Krasner, 2016), until in 1959 the model of the European Organisation for Nuclear Research (CERN) gave an impetus for the establishment of a European cooperative effort in space policy, leading to the foundation of the European Launcher Development Organisation (ELDO) and the European Space Research Organisation (ESRO), until in 1975 ELDO and ESRO were merged and ESA was created. Since then ESA has been responsible for coordinating European space activities.

The Single European Act (SEA) of 1986s is usually taken as a starting point marking the EC's/EU's development into a space actor, since it added research and development to the Community's competences (Madders & Thiebaut, 1992; Reuter, 2007, p. 12). Before that, the Commission of the European Communities has already participated as an observer in the European Space Conference (ESC) since 1970 (Hobe, Kunzmann, & Reuter, 2006, p. 64; Madders & Thiebaut, 1992). The ESC was the dominant forum for Europe's space community from 1967 to 1975 (Madders, 1997, p. 149), with the objective to harmonise space policies in Europe (Sheehan, 2007, p. 78). Since the SEA, the EC/EU has gradually increased its interest in space activities culminating in the inclusion of space as a shared competence between the EU and the Member States through Article 189 of the Lisbon Treaty of 2009, thereby creating a legal basis for the EU's action in the space field. Member States can thus choose which "organisation" to use for what space projects. The cooperation of EU and ESA is based on the so-called FA of 2004, which introduced the "Space Council", comprising both the EU Council of Ministers in form of the Competitiveness Council and the ESA Council. The introduction of the EU's flagship programmes Galileo, the European Navigation Programme, and Copernicus, the Global Monitoring for

N.-L. Remuss, *Theorising Institutional Change: The Impact of the European Integration Process on the Development of Space Activities in Europe*, Springer Theses, https://doi.org/10.1007/978-3-319-95978-8_3

Environment and Security programme, show the EU's commitment to the space area. The EU's ambitions to incorporate space policy into community policies, raise questions concerning the future of ESA in light of this EU aspiration.

Today's European space landscape is thus divided into three distinct levels:

- the EU;
- the intergovernmental organisations ESA and EUMETSAT; and
- the national space agencies.

In fact, the institutional problem of setting-up a structure for a coherent European space policy has already been recognised in the early 90s (cf. Schrogl, 1993; Hutter, 1991), but the question of the institutional arrangement is still not solved today. The FA (Council of the European Union, 2003) clarifies (inter alia) the purpose, principles, and fields of cooperation between the two organisations, but does not cover in detail the institutional arrangements. With EU and ESA both acting in the field of European space activities with differing but substantially "dual memberships" and partially overlapping competencies, questions regarding coherence (Rüttgers, 1989, p. 127), closer cooperation (Hobe et al., 2006, p. ix) and the adaptability particularly with regard to the institutional arrangement, gave rise to an academic and political debate about the future institutional set-up and division of labour between the EU and ESA (cf. for example Hörber, 2016a, p. 53). In this context four scenarios for an institutional set-up or a division of labour between EU and ESA have been debated:

- the cooperation model; i.e. improved cooperation between the two institutions
- the implementing versus political actor model; i.e. differentiation between ESA as the implementing and EU as the political actor
- the integration model; i.e. integration of the EU Commission in ESA's institutional structure and
- the accession model, i.e. the accession of ESA as an agency in the EU's institutional framework (e.g. Madders & Thiebaut, 1992; Reuter, 2007, p. 12; Hobe et al., 2006, p. ix; Kries, 2003; Hobe, Heinrich, Kerner, & Froehlich, 2009; Hörber, 2009a).

The European Commission itself has eventually tasked the independent consultancy Robert Berger in 2013 to analyse the evolution of the EU-ESA relations, providing a cost-benefit assessment, with the objective to propose several future options. Similar to the options already proposed by several experts, Berger summarised the following four options:

- Option 1: Baseline scenario—No action is taken
- Option 2: Improved cooperation under the "status quo"—ESA remains an independent intergovernmental organisation and there is a revision of the existing 2004 EU/ESA FA
- Option 3: ESA remains an independent intergovernmental organisation within which new structures are created to deal with EU programmes in accordance with the EU financial regulation and laws. These structures within ESA would be accountable to the EU; an EU-Chamber is foreseen staffed by the EU Commission
- Option 4: Transforming ESA into an EU agency while preserving ESA's successful intergovernmental features. (European Commission, 2012, p. 4).

Even though ESA's Member States have agreed to keep ESA as an intergovernmental organisation during the ESA Ministerial Council of 2014, the discussion about ESA becoming part of the EU framework continues. Europe has developed beyond the guiding ideals of European integration for the purpose of peace and prosperity after the Second World War and has now entered a phase of consolidation and exploration (Carpenter, 2016, p. 239). Against this background, the "hot potato" currently under debate among experts in the space community is: "What will happen to ESA?". Since academics are no prognosticators, this dissertation cannot anticipate future developments. Alternatively, this dissertation conducts a historic analysis of the development of the inter-institutional relationship between the EU and ESA, guided by the question: **To what extent has ESA changed institutionally so far?**

ESA is not the only institution, becoming questioned, as the EU extends its policy field. The evolution of the EU as an international actor, integrating ever more policy fields, rather led to a process of institutional change of several already existing international organisations. Some of these, were included as EU agencies (the WEU Satellite Centre (cf. Davara, 2000), the WEU Security Institute (cf. Council of the European Union, 2006) and the WEAO & WEAG) into the community framework, others were dissolved (the WEU, the JAA), and a third group of institutional actors managed to co-exist (CERN, CEPT). Thus, the case of the inter-institutional relationship between the EU and ESA might give interesting insights into the following question: "**What is the impact of the European integration process on existing institutional actors?**". **When, where and how are the EU's supranational institutions such as the Commission able to implement their own agenda?**

To answer this question, this chapter provides an updated overview of the development of space policy in Europe to qualify to what extent ESA has changed institutionally in a subsequent fourth chapter. The development of space activities in Europe is considered a case study of institutional change. In line with Campbell (2004, p. 40) if we want to determine, how much institutional change has occurred in a particular setting, we need to examine all the relevant institutional dimensions over time. Thus, the development of European space activities is considered from the beginning of European space collaboration in 1959 until 2014. This is in line with Dudley and Richardson (1999), who perceive long-term historical policy studies as the only way to understand the development of institutional change (Hörber, 2016b, p. 254). Moreover, in line with Thelen (1999, p. 400); Hall and Taylor (1996a); Conner (2010, p. iiix), the process of institutional foundation and selection can be linked to the process of institutional change. Accordingly, knowing how institutions were constructed "provides insights into how they might come apart" (Thelen, 1999, p. 400).

Space policy is a particularly interesting case. First, ESA was founded as an independent institution entirely separate from the European integration process that resulted in the EU. Thus, as Hörber (2016) put it, "the fact that there is now close cooperation with the EU is significant and worthy of investigation" (Hörber, 2016a, p. 54). Secondly, space policy has the potential to foster what European integration has promised: peace and prosperity, while it will always have the potential to attract the mind and hearts of many (Hörber, 2016, p. xxii). Directly or indirectly, space is relevant to every EU policy area (Hörber, 2016c, p. 11). Space is very much at the cutting edge of current EU policy developments. Recent phenomena such as climate change, rapid developments in crisis management and prevention etc. all need space capabilities to "manage" them (Ibid.). As Hörber (2016b) mentioned, space policy itself can also be considered a driving force of European integration. As such it has not yet been sufficiently understood because "space has not been studied as a policy domain as have defence, environment, transport or climate change." (p. 254) The strategic potential of space for the EU is likely to grow, given the many possible future fields of engagements such as commercial launches, missions to Mars, Moon research or a Moon village. The industrial and economic potential of this area of activity is a further motivation to strengthen the common frame between the ESA and the EU (Hörber, 2016a, p. 61). Moreover, space policy has the potential to foster a European identity (cf. Rathgeber & Remuss, 2009; Remuss, 2011).

As it has been outlined already, the time frame over which institutional change is examined is important (Aminzade, 1992). Thus, a chronological approach has been chosen as to identify changes. Special attention is paid to the impact of the European integration process on existing institutional actors.

To structure the observations of institutional change, the previously proposed stages of institutional development and facets of institutional change will be applied to the case study in a subsequent chapter. To do so, the historic development has been divided into several distinct phases, which, in line with Aminzade (1992) and Campbell (2004, pp. 44–6) can be distinguished by critical events. The following six phases have been identified:

1. Phase 1: The historic evolution of ESA (1959–1974)
2. Phase 2: From the founding of ESA to the entering of the EC in the space domain (1975–1986)
3. Phase 3: From the first steps of the EC to the ESA-EU Framework Agreement (1986–2004)
4. Phase 4: From the Framework Agreement to the Lisbon Treaty (2004–2009)
5. Phase 5: Post-Lisbon Treaty developments (2009–2012)
6. Phase 6: Post-ESA-EU Declaration of November 2012 (Post-2012).

Phase 1 and 2 can be separated by the foundation of ESA, which can be considered to be a critical event. Phase 2 and 3 are separated by the EU's first steps to enter the space arena through the introduction of research and development policy as Community competence by the SEA of 1986, which many observers (Madders & Thiebaut, 1992; Reuter, 2007, p. 12) of the development of the EC and today's EU to a space actor mark as the starting point. This development led to the formulation of the FA between the EU and ESA in October 2003, which provided "a common basis and appropriate operational arrangements for efficient and mutually beneficial cooperation" (Battrick, 2004, p. 29) between ESA and the Commission. It can be considered another critical event separating Phase 3 and 4. The FA resulted in the introduction of the common institution "Space Council", which characterise the fourth phase. The continuous rapprochement of the European Commission resulted in space being included as a shared competence in the Lisbon Treaty, signed in 2007 and entering into force in 2009. This critical event marks the separation between the fourth and the fifth phase. Following the Lisbon Treaty ESA and the EU were ever more struggling as to define their roles in the space policy field. ESA Member States thus considered the topic as part of their ESA Ministerial Conference in 2012 and issued a declaration entitled "Political Declaration towards the European Space Agency that best serves Europe" (ESA, 2012). This can be considered another critical mark, as ESA for the first time dealt publicly with the issue. It separates the fifth and sixth phase.

By taking these phases as a structure, this chapter will outline the development of space activities in Europe from 1957 until 2014.

3.1 Phase 1: Historic Evolution of ESA (1959–1974)

Phase 1 was mainly characterised by the first steps of the European States to cooperate in space matters. It covers the pre-ESA phase with ELDO and ESRO being founded as international organisations to deal with space in Europe. In Europe space activities have developed independent of the general process of European integration (Silvestri, 2003, p. 11). As a result, the European space system of today is a complex of "institutions, artefacts, national and international networks, production facilities and commercial activities" (Krige, 1994a, p. 1). This is why it is often referred to as "a tale of several Europes" (Madders, 1997, p. 1).

3.1.1 Background: The Situation in Europe

Until 1959, individual States exclusively conducted space policy. Especially for the U.S. and the Soviet Union, space was an important field of political and military confrontation, demonstrating their superiority. The two States took the lead not only in political and military matters but also in space affairs (Lüst, 1989, p. 147). The space race was mainly between the Soviet Union and the United States. Considerations of prestige and power characterised and drove their programmes.

In Europe space efforts were conducted on a national basis only, with the UK being the leading country. The UK had already worked on the development of solid-fuel guided rockets during the Second World War, resulting in the launch of three captures V-2s with the help of German specialists in October 1945. Since 1955 they were collaborating with the U.S. to develop an IRBM, called Blue Streak.

Second to Britain came France, where a group of German rocketeers had settled in 1946–47, forming a nucleus of the first French rocket teams. In 1949, the government had set up a specific laboratory for the research of aerodynamics and ballistic missiles. Regardless the initial budget limitations the laboratory was able to develop the Veronique rocket, modelled on the German V-2. Despite this huge achievement, the French effort "tended to limp along" until 1958, when the launch of Sputnik and the arrival of General Charles de Gaulle in power, drew attention to the area. It was de Gaulle's determination to develop an independent nuclear capability, which gave an enormous boost to the rocket development in France. The result was i.e. a strong governmental support for scientific and technological research at the time.

Italy too had an important national space programme, stimulated by the military. The Air Force supported by German expert Oberth collaborated with various companies, developing rockets and in 1959 launched a programme for upper atmosphere research. A launch pad was also established in Sardinia. In 1962 Italy accepted an offer by NASA to collaborate in space research, resulting in the first Italian satellite, San Marco-1, which was launched on 15 December 1964.

The Federal Republic of Germany at the time had troubles reviving its national space programme. Regardless the increasing pressure for the development of national space efforts in the 1950s, the V-2 weapon had damaged the public image of space in the country and the restrictions imposed by the allied powers did not allow any activity in rocket technology. As a consequence, interest groups in Germany were very receptive to the initiatives taken at the end of 1959 to launch a collaborative European space programme, considering it both as a legitimisation of Germany's re-entry into the field of space research, and as a platform from which to launch an independent national programme.

The above-mentioned national efforts in Europe could not compete with the space programmes developed in the U.S. and the Soviet Union, with the super powers being motivated by strong political, ideological and military interests. They were prepared to spend money, intellectual resources and industrial capability for the pursuit of their goals.[1]

3.1.2 First Steps Towards a European Space Collaboration

It was the success of the European Organization for Nuclear Research (CERN) that gave rise to calls for collaboration in the area of space (Peter, 2007, p. 99; Lüst, 1987, p. 8). CERN's everyday working had produced a genuinely "European" scientific community in this field. European scientists feared to become reviewers of literature and not producers of science in the fundamental area of space research. One way out of this was to partner with NASA in cooperative projects under governmental arrangements. Another possibility was to establish an overall European space science programme as proposed by the U.S., who envisaged a European organisation attached to NASA under the framework of NATO. Favouring a civil approach, the European Member States rejected this approach given NATO's military role (Madders, 1997, p. 29).

The first call for the "creation of a European centre for rocket research" managed by scientists on the model of CERN dates back to an unsigned article by J. Blamont, which appeared in "Les cahiers de la République" in November/December 1957. In

[1]This section is largely based on Krige and Russo (2000, p. 9–11).

1958/1959 NATO also started to actively promote the idea. Already in 1957 it had set up a task force on scientific and technical cooperation, suggesting the organisation of a space research programme and establishing a European NASA (Suzuki, 2003, p. 45). Several governments had already taken steps as well to establish national space committees in the late 1950s, since space exploration was perceived as "a symbol of scientific and technological prowess, an index of political power and a component of military strategy." (Krige, 1992, p. 8) By the end of 1959 when CERN's proton synchrotron reached its design energy of 25 GeV for the first time, it was clear that European scientific and technical cooperation could work, and provisions were being made to create a European atomic energy organisation within the framework of the negotiations leading to the setting up of the Common Market (Krige, 1993, p. 4). It was thus natural to consider extending it to the new and challenging field of space.[2]

The two main actors in the founding of the ESRO were the two scientists Edoardo Amaldi and Pierre Auger, who were also key actors in the process which led to the foundation of CERN. They became known as the "founding fathers" of the European space effort. Their first contact seems to go back to a letter from Amaldi of 6th February 1959, followed by a first meeting in Paris two months later, where the creation of a European organisation for space research was first discussed (Krige, 1992, pp. 1–2). The timing was not coincidental. As the International Geophysical Year was drawing to a close, Amaldi and Auger could exploit the linkages that had been established between scientists during this very successful initiative. Additionally, European governments were more than likely to welcome such a move in the light of the experience with CERN (Krige, 1993, p. 4).

Previously, in 1958, Amaldi had already discussed his dream of "Eurolune before 1965", the launch of a Euromoon from a European space organisation, with several experts among them Luigi Crocco, Professor of Aerospace Propulsion at Princeton University, Luigi Broglio, Italian expert in rocket technology, Isidor I. Rabi, one of the most important U.S. physicists at the time, Harrie Massey,[3] space scientist and scientific statesman in the UK and soon-to-be chairman of the British National Committee for Space Research, as well as Francis Perrin, one of the most influential leaders of the French physics community (Krige & Russo, 2000, pp. 13–4). Based on these first discussions, Amaldi worded what he thought should be the main feature of a future European organisation for space research.[4] He envisaged it to be modelled to CERN. In the aftermath, he continued trying to find supporters for his idea, but this first round of consultations with colleagues "resulted in an almost unanimous chorus of warnings against a straightforward application of the CERN model to space" (Krige & Russo, 2000, p. 18). It was only until he met Auger that his idea worked out.

---

[2]This section is based on Krige (1992, pp. 6–9).

[3]For a detailed discussion on the role of Britain in the development of European space activities cf. Massey (1986).

[4]For more information on the details of this letter refer to Krige and Russo (2000, p. 15). They are not in detail relevant for the analysis.

Amaldi first called for an independent organisation for conducting space research activities through an article of the December 1959's issue of L'expansion de la Recherche Scientifique. Therein, he proposed a small organisation similar to CERN, which he foresaw as operational by 1961. Madders (1997) refers to this move as a "carefully orchestrated affair", given Amaldi's discussion with Auger, "himself also a "leading organiser" and participant in CERN" (Madders, 1997, p. 30). On national level Auger was the head of the Comité français de recherches spatiales and took part in the planning of the French space research programme that was approved in the same year. Additionally, Auger in his function as Director of Sciences at UNESCO in Paris, together with Amaldi engaged in the politics of international space science during the process that led to the creation of the Paris-based Committee on Space Research (COSPAR).[5]

Shortly thereafter, Amaldi drafted a document entitled "Space Research in Europe", dated 30 April 1959, which he circulated among the senior science administrators of Belgium, France, Germany, Italy and The Netherlands as well as to CERN's Director-General and to the President of the EURATOM Commission, to encourage other European institutions to take an interest in this initiative. Therein Amaldi underlines that Europe could only compete with the U.S. and the USSR in this important scientific and technical field, when working together (Lüst, 1987, p. 8). He stressed the urgency for Europe to engage in space matters. Accordingly, there was a danger that European space research was "destined to remain a monopoly of the U.S. and the Soviet Union" with the countries of Europe being "mere spectators for the grand endeavours to the East and West of our continent. As a solution to this problem he saw the creation of an international organisation which would be pooling the resources of the European countries, thereby enabling the European scientist to make a valuable contribution to the exploration and study of outer space. Accordingly, if a European organisation or "at least a fairly well-founded provisional precursor of it", could be operational before the end of 1960, Europe could hope to close the gap between "herself and the Soviet Union and the United States before 1970." (Krige, 1992, p. 4) Amaldi made clear that he foresaw a research organisation, independent of any kind of military organisation, claiming that "If the military maintained a monopoly on the construction of rockets (…) each European country would build its own" (Krige, 1992, p. 5). Thus, Amaldi stressed to take CERN as a model. He envisaged the organisation to be free from bureaucratic and political pressures, with scientists having the power to shape the programme(s).[6] Krige (1997, 1992) suggests that Auger and Amaldi were initially torn between the model of CERN (intergovernmental model) and conducting space research under the European Communities (EURATOM), proposing "EUROLUNE" as a name for a new space organisation. EURATOM was eventually however decided to be too much subject to political contingencies to act as a model for a space organisation (Madders, 1997, p. 30; Krige, 1992, pp. 7–8). Additionally, the EURATOM model might have been inacceptable for the British. As mentioned earlier a "European NASA" as proposed by NATO, was

[5]This section is based on Madders (1997, p. 30).

[6]This section is based on Krige (1992, pp. 2–7).

not acceptable by the European States, as this would imply commitment to military space activities, which was not foreseen (Suzuki, 2003, pp. 44–5).

Auger first discussed the "possibility of creating a European space institute, quoting CERN as an example of success" (Krige, 1992, p. 10) at the first General Assembly of COSPAR, in Nice in January 1960. COSPAR can be said to have been a spin-off of the International Geophysical Year.[7] It allowed Auger to organise European space scientists (Madders, 1997, p. 30), who felt that European countries were so far reduced to the role of spectator compared to the progress of both the U.S and the USSR. The Amaldi-Auger proposal was supported by Massey, the president of the British National Committee for Space Research (Suzuki, 2003, p. 41), who not only wished to be fully involved in the discussions (Madders, 1997, pp. 30–1),[8] but also "swung the initiative towards Britain" (Ibid., p. 31).

The involvement of Massey however brought new light to the problem of launcher. Amaldi and Auger saw "getting access to a launcher" as "the critical problem inhibiting European space science activities" (Suzuki, 2003, p. 41). It was Massey, who proposed to use the British intermediate range ballistic missile "Blue Streak" as a European launcher (Ibid.). According to the UK Blue Streak was not only readily available but it "could [also] turn Europe into a participant in, rather than a spectator of, civil space activities" (Madders, 1997, p. 21). Massey proposed a formal but still non-governmental conference for scientists from CERN States as the next move (Madders, 1997, p. 31).

In the following, eight scientists from eight different countries met on 29 February 1960 in Paris in Auger's flat, where they proved their interests in a joint European space research effort. Most strikingly, the British were most enthusiastic about European space collaboration (Krige, 1992, pp. 10–11) and "Massey took the initiative in raising the idea of a formal negotiation with governments" (Suzuki, 2003, p. 41).

This private meeting was followed by another informal one, which took place in the rooms of the Royal Society in London on 29th April 1960, where about twenty European space research scientists from ten West European countries reported on their national activities and "identified three main areas which countries could profitably make use of one another's existing facilities" (Krige, 1992, p. 11): (1) exchange of visiting scientist between universities and government funded institutions; (2) coordination of satellite tracking on a European level; and (3) cooperation in sounding-rocket experiments using launching facilities in geographically different regions (Ibid., p. 12). Additionally, the British proposed to cooperate in artificial satellite experiments. Their plans in this area were well advanced. A design study had been in progress for six months and the British scientists hoped to be able to place design contracts by the end of 1961. The meeting took place at an opportune time, with Blue Streak's formal cancelation as a missile only a fortnight before

[7]The International Geophysical Year (IGY) in 1957 was proposed by the US government with the objective to establish global cooperation in space research through special IGY events (Suzuki, 2003, pp. 40–1).

[8]This is particularly striking as Amaldi's previous article in L'Expansion de la Recherche Scientifique was warmly welcomed by European science figures except by the British.

(Madders, 1997, p. 31). The British thus informed the other nations present that their government was possibly going to cease the development of its rocket and proposed a European research programme based on the use of Blue Streak as a launcher. The proposal was well received with various delegates making suggestions regarding the kind and level of contribution they could envisage for a joint European programme (Krige, 1992, pp. 12–3). It was finally agreed that Auger should call another meeting within two months of delegates "formally nominated by their national committees", and "empowered to create a Preparatory Committee for the establishment of plans for an extended European collaboration in space research." (Ibid., p. 15)

The next meeting took place in Paris on 23rd and 24th June 1960, where Auger proposed the "Draft Agreement Creating a Preparatory Commission for European Collaboration in the Field of Space Research." Given however that the scientific programme of any envisaged European joint venture was not clear, the question of the membership of the preparatory Commission and particularly of Australia was not clarified and those present did not have the authority to take decisions which would be binding on their governments, the preparatory commission could not be set up at that meeting. Thus, a study group, The European Space Research Study Group—(GEERS) was established with the task of clarifying these three points with Auger as its executive secretary (Madders, 1997, p. 32). The French government offered to host the bureau while the Swiss agreed to convene the intergovernmental meeting. A drafting committee modified the original paper prepared by Auger und met on 5 July 1960, where a new three-page draft agreement establishing the preparatory commission was summarised. Apart from the procedural questions, the main point of discussion at this meeting was the composition of technical study groups. While France and Britain had most expertise in the development of launchers several other countries, such as Italy, the Netherlands and Sweden wanted to be involved from the very beginning to gain the necessary experience in space research, which until then had been a monopoly of the larger European countries (Krige, 1992, pp. 20–1). The representatives of the smaller countries favoured the U.S. launcher as it "would cost half as much as developing the Blue Streak, but the British and French representatives insisted that the U.S. launcher (SCOUT) would be too small and unsatisfactory" (Suzuki, 2003, p. 46). At the same time, the smaller countries were apprehensive "that the big countries (…) would (…) swallow up a large part of the space budget", thereby threatening their autonomy (Ibid.). GEERS showed the different views of the countries regarding the aims of the organisations (Madders, 1997, p. 32). Given the problem of GEERS to overcome the difficult question of pertaining the development of launchers and autonomy of smaller countries, the question of launcher development was taken up by national governments under British initiative (Suzuki, 2003, pp. 46–7). Thus, in parallel to the initiatives taken by the scientists, high-level negotiations took place between European governments on the question of launchers. The main actors at these talks were Britain and France with at least "Belgium keeping a very close eye on the developments" (Krige, 1992, p. 22). France hoped to draw Britain close to the continent both politically and economically (Krige, 1993, p. 14).

Preparations for the intergovernmental meeting scheduled to take place in Switzerland, continued in London in October 1960, where Auger put together the previous documents and reports. His synthesis was divided into four main sections: (1) scientific, technical, economic, and indirect benefits to European cooperation in space research, (2) possible scientific programmes, (3) general principles and organisational structure envisaged for the new organisation, and (4) the activities to be undertaken by the preparatory commission. Additionally, the underlying philosophy and possible organisational structure of the new European agency was defined. The organisation was envisaged to be involved "in all stages of space research form securing vehicles and using launching sites to the exploitation of the scientific results and the processing of the data". Its international character was to be "reflected in the geographical distribution of its establishments, in the composition of its scientific and technical personnel, and in the allocation of contracts to industry" (Krige & Russo, 2000, p. 34). Krige (1992, pp. 23, 25) highlights two points about this meeting: (1) the determination of the scientists to control, as far as possible, the new European space research organisation, and (2) the fact that the delegates did not discuss (as they had in April) any collaboration in the development of Blue Streak as a European launcher. Blue Streak was avoided on the one hand due to the high-level discussions, which were still not finalised, and on the other hand developing a European launcher capability on the basis of Blue Streak was only treated as one option among others. Already then, technical experts were beginning to distance themselves from the issue, to consider alternative ways of achieving their scientific objectives and to narrow the scope of the envisaged organisation accordingly.

3.1.3 Institutional Bifurcation

Debates of such fundamental points related to the organisation's scientific role, Australia's membership, and the launcher development far exceeded GEER's informal role. Thus, based on GEER's report, a governmental conference was convened from 28th November to 1st December 1960 in Meyrin to review which further direction the initiative should take and to establish a governmental preparatory commission (Madders, 1997, p. 32). Twelve countries finally participated in the work of the COPERS: Austria (joined in October 1961 but withdrew later), Belgium, Denmark, France, the Federal Republic of Germany, Italy, The Netherlands, Norway (withdrew in June 1962, Spain, Sweden, Switzerland and the UK (Krige, 1993, p. 12). The so-called "Meyrin" agreement established the European Preparatory Commission for Space Research (COPERS), which was tasked with creating the framework for a "pure" European Space ***Research*** Organisation (ESRO) [emphasis added] (Lüst, 1987, p. 8; Krige, 1993b, p. 1), was opened for signature. COPERS was structured in a way that left Massey and Auger with considerable authority. Under its agreement, a three-quarters majority was needed for all substantive question but a so-called "bureau" consisting of the Commission's President, Massey, and his two Vice-Presidents (Prof. Broglio and Prof. van der Hulst) could exercise all powers delegated to it by the Commission between COPERS sessions. The Executive Secretary, Auger, was tasked with the implementation of COPERS' decisions but also to take decision "which it entrusts to him". As a consequence, he shaped the technical and scientific philosophy for ESRO and became ESRO's Director General (Madders, 1997, pp. 33, 37).

The Meyrin Conference was also, as Madders (1997) put it, "an orchestrated affair" (p. 32). Taking place in Switzerland, there was an unmistakable hint as to which way things should go (Ibid.). Participating governments started to question the narrow objectives of the physics' community thereby threatening the scientists' analogy to CERN. In contrast to CERN's particle accelerator, which cannot be adapted to another purpose other than highly specific research, rocket and satellite technology is adaptable. In fact, the question of the development of launchers was not considered by this conference, as at the start of the meeting the British as well as the French delegation made it clear, that the conference should not discuss this matter. Accordingly, it was feasible to create a European organisation dedicated solely to scientific research, which was not concerned with commercial applications such as telecommunication or with the construction and development of rockets (Krige, 1993, p. 11). As a consequence the option for a unified organisation had little chance. Particularly, the scientists' powerful backer—the UK—determined to split the space field by creating two institutions: one for launchers and one for space sciences. While, Massey appealed for space research to be left free of government interference, the British and the French governments did not want scientists to interfere with their strategic technology cooperation. The Anglo-French launcher cooperation acted before a defence to the two-organisation approach could be organised, thereby pre-empting any inquiry into the implications attached to the institutional separation (Madders, 1997, pp. 32–3).

This development can be seen as a turning point in the process of institutional foundation (Suzuki, 2003, p. 47). In contrast to Amaldi's and Auger's initial vision, it became clear that Europe would enter space with not one, but two organisations (Krige, 1993, p. 12). It was agreed in 1960 to found two space organisations, one similar to CERN dedicated to basic research in space sciences named ESRO and the other one to the development of launchers named ELDO (Krige, 1992, pp. 1–2: Krige, 1993b, pp. 1–2).

The agreement on the establishment of COPERS (see above) was overshadowed by the development on the question of launchers. Several technical and ministerial exchanges took place between France and Britain in 1960, discussing the potential involvement of France in the development of a rocket using Blue Streak as a first stage and Black Knight as a second. While France

> was willing to associate itself with the British government in a proposal to other European States that a study be made of the technical and financial possibilities of building in Europe a rocket system able to put heavy satellites in orbit (Krige, 1992, p. 28),

it was a necessary condition for any such collaboration that the second stage of the rocket would be built in France, and not be Black Knight (Ibid.). This principle of "national stages" ignored however the relatively advanced stage of development of

Blue Streak "and must be contrasted with allocation of work according to technical merit, cost or another independent criterion" (Madders, 1997, p. 23). To sum up, the French were cautious not about studying the possibilities of producing a European system of rockets, but they had serious technical and financial doubts about Britain's proposal (Krige, 1993c, p. 12).

3.1.4 The Founding of ESRO

COPERS held its first session in Paris on 13th and 14th March 1961.

> Its first task was to create the organs needed to define the scientific programme and necessary infrastructural facilities of the envisaged organisation, to draw up its budget and to prepare a convention for signature by those Member States' governments, who wished to join it. (Krige, 1993b, p. 3)

It established a working group on scientific and technical matters (STWG) and one on legal, administrative and financial ones (LAFWG). Their task was to define ESRO's mandate in details (Madders, 1997, p. 34). Europeans consciously decided not to pursue the possibilities of the new organisation being sponsored by the UN, or even by the Organization for European Economic Cooperation, because this would allow non-European participation (the U.S. and Canada were about to join the OEEC) and reduce its impact as a vehicle for European cooperation. Important was the fact that Switzerland "took the lead in hosting the key international meeting of government representatives that created the working groups, who drafted the founding documents for the proposed organisation" (Sheehan, 2007, p. 77).

The ESRO Convention and associated protocols were signed at a conference held on 14 June 1962.[9] Its convention entered into force on 20 March 1964 with the founding States being Britain, France, (West) Germany, Italy, Belgium, the Netherlands, Sweden, Denmark, Spain and Switzerland. Austria and Norway had observer status. According to its Convention, "ESRO was to promote collaboration among European States exclusively for peaceful purposes" (Ibid., p. 81). It focussed on launching small sounding rockets to the upper atmosphere and small satellites to near-Earth orbit (Ibid.).

Within the first four years, a number of facilities have been established, which were required for the European space programme to operate effectively. According to Krige (1993), the scientists setting up ESRO did agree more or less on the establishments and site choices. The main sites were: ESTEC, the European Space Technology Centre, originally called the Payload Engineering Unit as the core of ESRO, ESRANGE; the facility carrying out a sounding rocket programme in the auroral zone, ESTRACK, a network of tracking and telemetry stations controlling the spacecraft once in orbit and serving as receptors for data transmission, ESDAC,

[9] For more details on these early developments and the drafting of the ESRO Convention cf. Krige (1993, p. 1).

the European Data Analysis Centre, predicting and analysing satellite orbits in anticipation of a launch as well as analysing and processing the data received through the tracking network once a satellite was operational, ESLAB, undertaking theoretical research of importance to space science and providing experimental facilities enabling individuals and small institutions to undertake research in space science and the headquarter hosting the Director General with his staff. These sites were eventually complemented by ESLAR (later renamed as ESRIN), a laboratory for advanced scientific research (Krige, 1993, pp. 28–30). On the basis of several bids for some of the facilities by several Member States, Odd Dahl, a Norwegian engineer who had also been active in launching CERN, was invited to make proposals for the geographical location of the sites (Sheehan, 2007, p. 81).[10] The budget, set at U.S $ 306 million, would be covered by annual contributions from Member States and be assessed on the basis of national income (Ibid.).

COPERS established two working groups, which were similar to that of CERN and consisted of a supreme governing body (the Council), which was advised by an administrative and finance committee (AFC) and a scientific and technical committee (STC/STWG). Recommendations for the STC were conducted by a launching programme advisory committee (LPAC/ LAFWG), with the task to define an appropriate scientific programme in the light of proposals it received from a number of working groups. Those members were specialised in different fields of space science. The STC and its subgroups defined a draft scientific programme and a launching schedule for ESRO's first eight years. Their proposals—the so-called blue book—were well received during the third session of COPERS held in October 1961 in Munich. The programme contained in the blue book was divided into short-term and long-term projects and has to be understood as a declaration of intent rather than a definite programme of work. It was intended to provide a rough basis for a first estimate of costs and a framework for priorities. The experimental work was divided into three categories depending on the source of funding, being pure ESRO experiments, combined ESRO and national experiments and purely national ones. In all three cases the construction of the scientific payload was under control of national groups instead of being controlled by in-house staff, since it was feared that ESRO would built up a strong in-house staff, thereby draining away the best brains form the national centres, which would rapidly become a privileged group with a monopoly over the best equipment and ready access to space on satellites (Krige, 1993, pp. 20–36).

While ESRO is largely perceived as having been relatively successful in fulfilling the objective, it was set up for: to encourage European collaboration in space research and technology, the financial resources committed in 1963 to cover an eight-year programme were rapidly shown to be hopelessly inadequate. Nonetheless, ESRO succeeded in managing to achieve a significant degree of autonomy in its organisation and in the implementation of its programmes. The 1968 (NASA-aided) launchings of the ESRO-I, ESRO II and HEOS I satellites heralded the arrival of Europe as a space power (Sheehan, 2007, pp. 81–2).

---

[10]For more information on these geographical choices, consult Krige (1993).

After a little more than two years, the Council found its agenda crowded by relative trivialities, and thus started to have doubts about the proper functioning of the established structure. At the same time the organisation's expenditure profile was tilting heavily in favour of administrative expenses. As a consequence, the Council set up a group of experts in 1966 to study the internal structure. Accordingly, ESRO was in a "crisis of confidence" due to an unclear distinction between legislative and executive parts of the organisation. For the AFC to get out of its blockage, it was proposed to lower the limits below where executives could award contracts without seeking the committee's approval. Additionally, the LPAC, it was suggested, was to put its proposals directly to the Directorate, rather than having them funnelled through the STC. It was then up to the Directorate to draw up a proposed programme for the organisation in consultation with the STC and the AFC, before laying it before the Council for final approval. Apart from this is was proposed that the STC should no longer be based on national representation but solely on recognised expertise in the scientific and technical aspects of the programme which the organisation was following at any given time. All other proposals were welcomed but this one (Krige, 1993, pp. 37–8). As when the organisation was founded, the Member States still thought this as their way to ensure that its interests were properly protected inside the organisation (Ibid., p. 93). The proposed redefinition of the decision-making structure for ESRO was an attempt to transfer the CERN model to the space organisation. The idea to delegate more authority to the Directorate mirrored the wish for "independent" scientific advice to be available through the STC and was part of the general desire to reduce the influence of the bureaucracies of Member States and their interest in the functioning of the organisation (Ibid., p. 39).[11]

Krige (1993, pp. 26–7) observed three characteristics of the founding community of ESRO. First, the European space science community was relatively young and inexperienced and was rather euphoric with the elements being put in place for their future organisation. Secondly, the leaders of this community did not want the new institution to compete with their national space programmes and were thus willing to see ESRO's expenditures being restricted. Third, NASA also offered to enter bilateral agreements for flying scientific satellites, which made ESRO one out of several options to achieve the objectives of the scientists.

3.1.5 The Founding of ELDO

The origins of ELDO can be traced back to mid-1950s, when Britain and the U.S. considered working together in the development of a long-range ballistic missile (Krige, 1993, pp. 12–3), which became known as Blue Streak (Krige, 1993c, p. 2), and France was completing the development of the Veronique rocket, capable of launching small satellites (Sheehan, 2007, p. 77). Two months after the Meyrin

[11] For a more detailed account on these developments refer to Krige (1994a, p. 2), Krige and Russo (2000).

conference Britain and France convened a meeting held at Strasbourg from 30th January to 2nd February 1961 and proposed to their European partners to collaborate in the setting up of an organisation devoted to the common development of a heavy satellite launcher (Krige, 1993, p. 12). The French attitude towards cooperation with Britain in this endeavour had suddenly[12] changed (Krige, 1993c, p. 16). An agreement was finally reached at this intergovernmental meeting. A programme was adopted for the development of a three-stage launcher with the first being built by the UK,[13] the second by France and the third, as well as a series of test satellites, by other Member States. It was accepted that France, Germany and Italy would contribute to the costs of the programme (Krige, 1992, p. 32). Participating governments were determined to maintain control over the shape of the launcher project and to define a form of collaboration, which protected their national interests. The scientists in turn feared that the launcher would be funded at the expense of their research, and realised that the political and military implications of rocket/missile development might impede collaboration with some of their colleagues from neutral countries (Krige, 1993b, pp. 1–2). There was a feeling that if Europe was to seek genuine autonomy in space then it clearly needed its own launch rocket (Sheehan, 2007, p. 78). A follow-up conference in November 1961, the so-called Lancaster House Conference,[14] led to the drafting of the Convention of the European Launcher Development Organisation (ELDO), which was signed in March 1962 by France, the UK, Italy, Belgium, West Germany, the Netherlands and Australia. (Australia's presence was necessary as the only available launch site was the Anglo-Australian testing and launch site at Woomera in Australia (Ibid., p. 77)).

ELDO's Convention obliges the Member States "to the development and operation of a space launcher and ancillary equipment" (Ibid.). It limited ELDO to only peaceful applications of the launchers and equipment. It was in 1966 that ELDO decided to begin construction of an equatorial launch site at Kourou in French Guinea because of the advantages of launching rockets from bases on the equator (Ibid., pp. 77–8).

As the European Economic Community (EEC) began to appear a successful initiative after 1957, the British government came to regret its initial decision not to participate in the organisation. Offering leadership in the field of European space cooperation was "one dimension of a wider strategy aimed at closer integration with the Six" (Ibid., p. 78) and was seen in the UK as an important proof of Britain's (new) European credentials. This perspective was echoed by West Germany, whose government welcomed the British proposal for a European launcher organisation as "an opportunity to strengthen the linkages of the UK with the continent as such and also as a possible first step towards and enlargement of the European Economic Community (EEC)" (Ibid.).

---

[12] According to Krige (1993c) there is no single or simple reason for this change. Perhaps the French always intended to join and were simply bargaining, which finally proved too difficult (Krige, 1993c, p. 16).

[13] For more information on the difficulties Britain was facing during the Strasbourg conference, refer to Krige (1993c, p. 19ff).

[14] For more details refer to Krige (1993, p. 28ff).

The most striking feature about the birth of ELDO, as acknowledged by Krige (1993, p. 19), is the scepticism and opposition to the project by many experts in the main participating countries. Both Italy and Germany[15] had to be persuaded in joining (Ibid., pp. 16–8). The main fear of the scientists was that the development of civilian launchers would be conducted at the expense of their national research programmes. Also, engineers perceived Blue Streak as already obsolete and thus saw little point in taking it as a basis. They regarded the division of the management and development of the four stages between the different countries as doomed to fail. These doubts were however overcome by both lobbying activities of the aerospace industries, which in September 1961 established a supranational body called Eurospace with the aim to "promote the development of aerospace activities in Western Europe", and by political consideration (Ibid., p. 19).

Britain (Thorneycroft) defended Blue Streak on the basis of four concerns: (1) to avoid U.S. dependency, particularly in telecommunications, (2) to benefit industry, (3) to enable Britain to take advantage of possible military applications of space and (4) to avoid criticism in Parliament for not having cancelled the venture earlier. Britain pressed ahead at all costs and thus committed itself to paying almost 40% of the budget of the new organisation by 1963 (Ibid., pp. 19–20).

3.1.6 ELDO's Demise

> [T]he conception, development and demise of the ELDO must contain a series of lessons, often painful, on how not to organise, fund, manage and operate a cooperative technical effort which crosses international boundaries. (Sheehan, 2007, p. 81)

Apart from ELDO experiencing severe budgetary problems, with Europe lacking experience in such programmes underestimating costs, the three-stage multinational rocket programme was not a success. It proved too difficult to combine the French and the British systems. Additionally, the German third stage was not a success (Ibid., pp. 78–9).

ELDO's difficulties were compounded by the policy of "juste retour" enshrined in the Convention. Under this philosophy, profits and jobs created through the programme were to be allotted to the Member States in proportion to the financial contribution they had made. However, with the Europa programme this was difficult to achieve, since the main elements had been allocated to a limited number of countries, and for reasons of efficiency it often made more sense to distribute contracts through competitive tendering rather than allocate them to a country that did not actually have the technological capability to deliver a cost-effective product. Disputes over the fairness of the way in which juste retour was being implemented damaged the working relationships between the Member States but were extremely difficult to overcome within the limitations of the programme (Ibid., p. 78).

---

[15]Without German support, the proposed organization would not be truly "European" cf. Madders 1997, p. 24).

The ELDO convention also did neither provide an institutional mechanism for political discussion within the organisation, nor was there an effective mechanism for coordinating the activities of ELDO and ESRO (Ibid., p. 78). This institutional separation of ESRO and ELDO proved to be impractical and rather hampered the deployment of competitive space activities. The two organisations lacked political coordination (Hobe et al., 2006, p. 538).

A fundamental political problem for ELDO was that it was caught up in the internal political tensions which bedevilled the other major European economic and security organisations in the mid-1960s: There was a profound political opposition between France and the UK on basic international political strategy (France wanted to achieve genuine autonomy in space for Europe, and saw avoidance of dependence on NASA and the U.S. as a crucial objective for a European collaborative programme. Britain in contrast wanted to utilise existing British technology to maximum effect, and for Europe to pull its weight in the North Atlantic Alliance and was opposed to any initiative that might weaken the links between the U.S. and ties European allies.). Britain believed a "wasteful duplication of U.S. technology was unnecessary" (Massey, 1986, p. 231) and that launchers could be purchased from NASA (Ibid.). West Germany in contrast saw the issue as a pragmatic one, and was happy to use NASA launchers, if that was the most cost-effective option. Indeed, it hinted that it might withdraw from ELDO if it could obtain NASA guarantees of launch facilities for national payloads (Sheehan, 2007, pp. 80–1).

The budgetary problems, together with the problems in developing the technology, led many of the Member States to reassess their commitment to the programme. The enshrined national stages principle[16] "however effective as political inducement" (Madders, 1997, p. 23) contributed to the failure and technical difficulties of ELDO. By 1965 the organisation was in crisis and an emergency conference had to be held (Sheehan, 2007, p. 78). The first ELDO conference of ministers in 1966 emphasised the need for close and effective coordination between the European space organisations and called for the creations of a European Space Conference (ESC) to gradually harmonise European space activities (Ibid.). The ESC became the dominant forum for Europe's space community from 1967 to 1975 (Madders, 1997, p. 149) and is also referred to as the political dimension of space (Mazurelle, Wouters, & Thiebaut, 2009, pp. 10–11). Later the ESC became a forum for crisis (Madders, 1997, p. 124). The first ESC took place in December 1966 and was attended by all the ELDO States plus Denmark, Spain, and Austria, Greece, Ireland, Sweden and Switzerland attending as observers (Sheehan, 2007, p. 78). It agreed on the ECS's role and passed a resolution "confirming the new conference's establishment and its objective of uniting Europe's space effort" (Madders, 1997, p. 125). The second ESC took place in July 1967. It was attended by all ELDO and all ESRO Member States. It was agreed

[16]The three stages of the European launcher were to be built by Britain (on the basis of Blue Streak), France (on the basis of Véronique) and Germany. The decision to do so was made on the basis of political considerations and not according to technical merit, cost or another independent criterion. While ignoring Blue Streak's advanced stage of development, the second stage's thrust and weight requirements became negotiable resulting in very demanding requirements places on the newcomer to rocketry—Germany—who built the third stage (Madders, 1997, p. 23).

to establish the ESC as a permanent body, which would meet once a year at ministerial level to work out and ensure implementation of a coordinated European space policy (Ibid., p. 79).

The critical interrelationship between European space policy and the wider politics of European integration was demonstrated in 1967, when President de Gaulle delivered another veto on Britain's application for EEC membership in November 1967. While for several years the UK, seeking to join the European Economic Community, had been playing the part of the "good European", the British government immediately started opposing French policy in several key areas, one of which was space policy, where the UK blocked French-led efforts to reform ELDO and significantly increase its budget (Ibid., p. 80). At the ELDO Ministerial Council Meeting in April 1966, the British delegation announced its discontent with the progress of ELDO, arguing that the gains were not equivalent to the British contribution. The change in their attitude can be accounted to the new policies initiated by the labour government under Harold Wilson. They were following a "white heat" policy; reassigning the cost-efficiency of technological investment and re-evaluating the high-cost and high-prestige programmes. At the same time, the British industrialist, frustrated about the performance and industrial and technological return, published a statement in May 1966 urging the government to move away from the priority of the launcher programme within ELDO towards a national one (Blue Streak-Black Arrow). As a result, the government decided to increase the funding of national programmes. Consequently, the British government announced at the ELDO Ministerial Council in July 1966 that it would only remain within the ELDO framework as the Convention prevented it from withdrawing and demanded a reduction of its financial contribution. As a consequence of the reduction of the British financial and moral commitment, the Italian worries about its industrial return (rather than of the success of launcher development) and the German retreatment to the background given the failure of the German-sponsored third stage, ELDO became more French-led. The French demand for a strengthened ELDO Secretariat was however not accepted by the other Member States.[17]

At the same time, there was a "crisis of confidence throughout the organization" (Krige, 1993, p. 72) in 1966 culminating in summer 1967 when the staff committees of ESTEC and ESLAB, in open revolt, addressed a note complaining about their working conditions to the Dutch Queen (Ibid.).

The tide however seemed to be turning when at a ministerial meeting in November the U.S government suggested for Europe to participate in the post-Apollo programme, and design studies for possible contributions to this were handed to ELDO. At the same time the last launch of the Europa I rocket experienced only minor technical difficulties. Against this background, the launch of the first Europa II rocket, which exploded shortly after the lift-off, proved to be a turning point, turning hope into disappointment and disarray (Krige, Russo, & Sebesta, 2000, p. 13).

[17]This section is based on Suzuki (2003, pp. 58–61).

3.1.7 Small Versus Big Countries

Some of the problems of ELDO were at the same time experienced by ESRO. In both cases the smaller countries (for ELDO: Italy, for ESRO: Spain) felt that their industries were not benefiting from the membership in relation of their contribution and thus became frustrated, feeling that their contributions went to the benefit of the already stronger and more competitive companies in Britain and France. In 1967 Spain threatened to withdraw from ESRO if their industrial return was not guaranteed (Suzuki, 2003, p. 63). The governments of the smaller Member States proposed introducing the principle of fair geographical return, or juste retour. Its introduction was on the agenda of the ESRO Council since 1965 but France and Britain were objecting the idea, arguing that it would undermine international competitiveness of European industry and that it is unacceptable to be "penalized for having competitive industry" (Suzuki, 2003, pp. 64–5). The conflict between bigger and smaller countries became deeper in the mid-1960s and was eventually solved by Herman Bondi's initiative to mediate the two sides. Under his lead the ESRO Member States found an agreement.[18]

Finally, in 1968 the British government announced that it would not continue to fund ELDO after 1971, justifying this move by declaring that all UK research programmes dependent on government support must meet the criterion of "economic justifiability" (Sheehan, 2007, p. 80). The question of British and Italian contribution to launcher development could finally be solved in the ELDO Ministerial Council Meeting in April 1969, with France, Germany, Belgium and The Netherlands agreeing to shoulder the shortfalls of the British and Italian share (Suzuki, 2003, pp. 69–70).

3.1.8 The ESRO Reform: Introduction of À La Carte Cooperation

European governments blamed the ways in which the European space programmes were managed for the crisis over European space collaboration. The idea that Europe needed a coherent space policy gained ground among the ESRO and ELDO Member States. To discuss such a policy, the Member States of ELDO, ESRO and the European Conference on Satellite Communications (CETS) formed the ESC in Paris in December 1966 (Ibid., p. 67).

First, Jean-Pierre Causse, the Director of Brétany Space Centre of CNES and the French space agency, was entrusted with the task of drafting a report on the options for European space collaboration. His subsequent Causse Report, submitted in December 1967, contained several ideas on how Europe should manage space programmes. First, science programmes itself were not sufficient reasons to jus-

[18]For details on the agreement refer to Suzuki (2003, p. 65).

tify the existence of ESRO and its expenditures. Thus, the inclusion of application satellites was proposed. Secondly the report stressed that application satellites and launch vehicles programmes were crucial for Europe to maintain international competitiveness. Thirdly, the report suggested Europe could develop direct broadcasting satellites. Fourth, it recommended the creation of a single European organisation for space. Causse himself held that launcher development had to be coordinated with satellite programmes and vice versa. It was inefficient to separate the two aspects in different organisations (Suzuki, 2003, p. 67). Also, the Committee held that some of the reasons for creating separated institutions were no longer valid and thus the consequences no longer desirable (Madders, 1997, p. 127). Causses's objective had been "to bring Europe into an age of substantial, autonomous exploitation of the space environment." (Ibid., p. 130). One of the clear difference of the approach of the Causse Report lay in the clear differentiation of the aims of European space efforts compared to the space programmes of the U.S: and the Soviet Union (Krige & Russo, 2000, p. 338). At the time the perception of civil space was dominated by the space race between the two superpowers and was thus "synonymous with international rivalry, national prestige, and man-in-space-programmes" (Ibid.). The Causse report emphasised the long-term commercial interest of space instead, as well as Europe's need to develop the collaborative autonomy to compete with the space giants: "Europe should above all demonstrate her determination to be independent. Rather than indulge in illusory competition, she should demonstrate her determination to be independent." (Ibid.).

In the following, the ESC created an ad hoc group chaired by Bannier, Dutch civil servant, who had previously chaired a commission to reform ESRO's internal organisation.[19] Relying upon the basic institutional approach of the Causse report, the Bannier group added "an elegant programme structure" (Madders, 1997, p. 134), designed to balance the need for solidarity and flexibility. It consisted of four tiers: basic programme, minimum programme, supplementary programmes, and special projects. A similar form of division was later adopted for ESA (omitting the basic programmes) but proved only as a springboard for discussion at the time of publication (Ibid., pp. 133–4). The Bannier group also considered the geographical dispersion of ESRO as detrimental to its proper functioning. As a consequence, a geographical reorganisation was proposed (Krige, 1993, pp. 41–2). Thus, in October 1968, a new ad hoc group, chaired by Spaey (Belgium) was tasked to review the institutions and the principles for participation in programmes. The Spaey group urged the creation of a single organisation "in which a degree of participation in a basic complex of programmes would be mandatory, with the exception of the launcher component" (Madders, 1997, p. 135). Introducing an à la carte system for European space collaboration, launcher development could be kept apart from other issue areas. Member States further agreed to form a single European space organisation for basic programmes with the option of opting out of the launcher development (Suzuki, 2003, p. 69). Heads of States also decided to investigate a possible merger of ESRO, ELDO and CETS (Sheehan, 2007, p. 82; Hobe et al., 2006, p. 538).

[19]For more information refer to Krige (1993, p. 37ff).

A fundamental crisis arose in 1970 when certain Member States strongly urged that ESRO's operations were to be extended from work on purely scientific satellite to include the development and launching of applications satellites, i.e. SatCom satellites launched for commercial rather than purely scientific purposes. This proposal had been hovering in the background for several years (Sheehan, 2007, p. 82).

At a meeting of the ESC in November 1970 the delegations from Belgium, France and Germany announced that they were determined to go ahead with a comprehensive space programme on their own (Krige, Russo, & Sebesta, 2000, p. 1).

During the first months of 1971 the ESRO Secretariat, set up a Steering Committee to define its own position. It prepared with Puppi a report, which was circulated among the delegations (Krige, Russo, & Sebesta, 2000, p. 2). On the basis of the Bad Godesberg meeting, the Puppi report submitted by the Committee of Senior Officials produced ideas to break the deadlock in the negotiations (Suzuki, 2003, p. 76). The report built on the earlier Causse, Bannier and Spaey proposals (Madders, 1997, p. 143). As introduced by the Dutch before, the report proposed Member States' participation on a case-by-case basis with scientific and research and development (R&D) programmes being mandatory. Secondly, it recommended European investment in addition to telecommunication satellites in application satellites (Suzuki, 2003, p. 76).

Additionally, the British delegation was reorganised. While it had previously consisted of scientist and bureaucrats, since mid-1971 the Ministry of Industry joined the delegation in response to the increasing importance of an application problem. The change in the structure of the British delegation also contributed to the process of transformation of the ESRO. Apart from the new members of the British delegation, the spirit of the delegation also changed. Given their joining to the EEC, a new political climate in Europe was reached. As a result, France, Germany, Britain and Italy agreed in July 1971 to begin studies for application programmes for navigation and meteorology alongside telecommunications. Given the introduction of the à la carte participation the unwillingness of Denmark to participate did not have repercussion in the negotiations (Suzuki, 2003, p. 76). In November 1972, the ESRO Convention was amended to include the development and launching of applications satellites (Sheehan, 2007, p. 82).

3.1.9 Towards a Single European Organisation

From as early as 1966, it has been agreed that Europe suffered from having too many space organisations and that some attempt should be made to coordinate the work of ESRO, ELDO and CETS (Krige, Russo, & Sebesta, 2000, p. 23). Interestingly, already then space has been considered of fundamental interest to the EC, as illustrated by Recommendation 130 of the Assembly of the Western European Union (WEU) adopted on 14 June 1966, which called for WEU governments to "[p]repare for the inclusion of a permanent European space vehicle launcher development organisation within the framework of a future single European Community."

(WEU Resolution cited in Mazurelle, Wouters, & Thiebaut, 2009, pp. 10–11; cf. Bailes, 2011). Apparently, policy-makers had considered the EC as Europe's space leader.

By 1970, ELDO had not managed to launch a single satellite with the Europa rocket, nor had it achieved anything, which would have been beyond the financial and technical capabilities of any of the larger Member States (Sheehan, 2007, p. 81), when the U.S. offered to European governments to participate in post-Apollo programmes (Madders, 1997, p. 145). Additionally, the blow-up of the Europa 2 in November 1971 disappointed all Member States. Particularly, Germany's enthusiasm for a European launcher was fading away in favour of the U.S. shuttle. Subsequently, the ELDO Council held in April 1973 decided to cancel Europa 2 because Germany declared it would only pay its rundown cost. The decision was confirmed by the ESC meeting in July 1973 (Suzuki, 2003, pp. 78–80). ELDO's activities ceased and its assets and staff were transferred to ESRO (Sheehan, 2007, p. 81; Krige, 1994a, p. 1).[20] The ESC set up a Committee to draft a single European Space Agency (Suzuki, 2003, p. 80). "Following almost a decade of tribulations of an "unmanageable ESRO-ELDO-ESC troika" (Mazurelle, Wouters, & Thiebaut, 2009, pp. 10–11), Member States decided to create a single entity, bringing together the mandates of all three organisations in the European Space Agency, signalling the first revolution in European space governance (Ibid.). The creation of ESA can be seen as the outcome of a hegemonisation process working in the field of European space organisation between European nations (Hörber, 2016d, p. 22).

3.2 Phase 2: From the Founding of ESA to the Entering of the EC in the Space Domain (1975–1986)

The second phase of the development of space activities in Europe is characterised mainly by the early years of ESA.

3.2.1 The Foundation of ESA

ESA, a new organisation, was established in 1975. The amended ESRO Convention formed the basis for its Convention after the December 1972 decision to set up the successor organisation (Sheehan, 2007, p. 82). It was signed in Paris on 30th May 1975 (Krige, 1994a, p. 1; Lüst, 1987, p. 9; Rüttgers, 1989, p. 127) by all ESRO and ELDO Member States[21] except Australia (Sheehan, 2007, p. 83). The signature of its Convention marked a "fresh start for Europe's space activities" (Madders, 1997,

[20]For further information on the history of ELDO consult De Maria (1993).

[21]Belgium, Denmark, France, Germany, Italy, the Netherlands, Spain, Sweden, Switzerland, the United Kingdom and Ireland.

p. 179). For some non-ESRO/ELDO Member States, the shift to a single European agency was a good opportunity to reconsider their membership. One such country was Ireland, which officially applied and whose membership was accepted in 1976 (Suzuki, 2003, p. 87). In 1978, Norway and Austria joined. Canada signed a cooperation agreement on 1st January 1979 and took part in several programmes. ESA inherited from ESRO a headquarters building in Paris, a payload engineering unit (ESTEC, the European Space Technology Centre) in Noordwijk—The Netherlands, a mission control centre (ESOC, the European Space Operations Centre) in Darmstadt, and a small documentation centre (ESRIN) in Frascati near Rome. It also took over a launch base in French Guyana and a globally dispersed satellite-tracking network (Krige, 1994a, p. 1). ESA consists of two principle bodies: a Council as the only decision-making body and an Executive led by the Director-General (DG) (Suzuki, 2003, pp. 93–4). Additionally, pursuant to Resolution No. 3 of the Final Act of the ESA's Council's first meeting, ESRO's Scientific and Technical Committee and administrative and Finance Committee were confirmed (Madders, 1997, p. 199).

The Preamble of the Convention gives insights into the reasons and objectives behind its foundation. Given

> the magnitude of the human, technical and financial resources required for space activities that lie beyond the means of any single European country (…) a single European space organisation is being established as to increase the efficiency of the total of European space efforts by making better use of the resources at present devoted to space and to define a European space programme for exclusively peaceful purposes. (ESA, 1975)

ESA shall "integrate the European national space programmes into a European space programme as far and as fast as reasonably possible" (Ibid.). It aims

> to provide for and to promote, for exclusively peaceful purposes, cooperation among European States in space research and technology and their space applications, with a view to their being used for scientific purposes and for operational space applications systems. (Ibid.)

The Convention foresees four purposes for ESA: long-term policy, activities, coordination and industrial policy (Suzuki, 2003, p. 87). The emphasis on industrial policy can be explained by reference to the U.S. and USSR advance. ESA's Member States wanted to create an industrial structure which could keep up with the U.S. and Soviet space programmes. A competitive European space industry was presupposed, which needed "some sort of safeguard for the investments to be taken in space research, technology, and development." (Micklitz & Reich, 1989, p. 27). As an incentive for the Member States to participate, the principle of geo-return was introduced, as to assure contribution to the "home industry". The importance of this principle can also be seen in the growing significance of the so-called "optional programmes" in opposition to the mandatory ones (Ibid.). As it can be seen, ESA was designed as an intergovernmental institution lacking political aspirations of any kind. It was foreseen as an effective executive agency of the European space programmes (Hörber, 2016d, p. 23).

To a large degree, ESA's Convention spelled out the rules and practices developed during the final years of ESRO and ELDO: The principle of juste retour and optional participation were strongly supported by the smaller countries, aiming at protecting their industries while the big countries became increasingly selective about their participation, favouring national programmes (Suzuki, 2003, p. 95; Madders, 1997, p. 214). ESA's Convention is just four articles longer than ESRO's and is composed in the "same terse style". In comparison to ESRO's Convention, which was "genuinely straightforward", the wording of ESA's Convention is "more nuanced, and articles interrelate to a much greater extent". Its key articles on purposes, programmes, the Council, facilities, information and cooperation saw major revision, and the articles on launchers and industrial policy are completely new. Protocols to the ESRO Convention now became annexes with few changes and thus an integral part.[22]

The international competitiveness of European space industry increased during the 1990s as the major companies began a series of mergers designed to keep them competitive in the face of similar developments in the post-Cold War U.S. aerospace industry. Additionally, in contrast to ELDO and ESRO, ESA developed long-term planning up to 20 years ahead and committed itself to a mix of large and smaller projects (Sheehan, 2007, p. 85). ESA's main characteristics are its industrial policy and the co-existence of mandatory and optional programmes. While the Member States are obliged to participate in basic activities, the scientific programme, and the funding of infrastructure, participation in all other, optional, programmes is voluntary. The success of ESA's optional programmes is to a great extent based on its industrial policy with the principle of geographical distribution the "juste retour" as its main feature (Hobe et al., 2006, p. 539).

The successful first launch of the "Ariane" rocket in December 1979 was crucial to the subsequent success of ESA. It gave Europe access to the growing market for launching applications satellites (Sheehan, 2007, p. 85) and proved that ESA was more successful than its predecessor ELDO.

3.2.2 Striving for Autonomy

The Ministerial Council held at Rome in January 1985 is often said to have marked a turning point for European space collaboration (Suzuki, 2003, p. 111). In Rome, the ESA Member States manifested both their intentions to develop a civilian system and their "commitment to maintain and develop independent European capabilities in space" (ESA, 1985). This commitment was driven by political and institutional reasons. The ESA Member States had learned the hard way, that an independent capability "was necessary to protect European projects from American unilateral positions" (Suzuki, 2003, p. 112). Previously ESA had signed a Memorandum of Understanding (MoU) with the U.S. concerning the Aerosat programme, which the Europeans regarded as binding when, the U.S. government decided to withdraw. This was something they wanted to avoid particularly in the context of the International Space Station (ISS). The commitment to the independent capability as well as a shared congeniality towards the U.S. attitude at the time of the invitation to Europe

[22]This section is largely based on Madders (1997, p. 180).

to participate in the ISS project produced a sense of identity among the participants at the Council meeting in Rome, which was described as the "spirit of Rome" (Ibid.).

A similar atmosphere must have been present at the Ministerial Council Meeting in The Hague in November 1987. The disastrous failure of the Challenger the year before was taken as a justification for developing an autonomous passenger system and the Europeans were prepared "to go it alone", should the U.S. use the ISS for military purposes (Ibid., p. 113).

3.2.3 *The Consolidation of ESA*

Then, ESA's first programmes had been or were being completed and national capabilities had been strengthened by both European and national programmes. ESA still remained the focal point, confronting space policy makers with a choice of going forward with intensified cooperation or returning to bilateralism. The prevailing geopolitical and European conditions including the need for a successor programme to Spacelab, encouraged adoption of the former option. The undertaking of manned space programmes fostered the idea of Europe becoming a major space power as part of its destiny (Madders, 1997, pp. 588–9). The invitation to participate in the ISS project was certainly an époque-making moment in the history of European space collaboration (Suzuki, 2003, p. 117).

3.3 Phase 3: From the First Steps of the EC to the ESA-EU Framework Agreement: Historic Evolution of Space Policy as a Community Policy Within the EU (1986–2004)

ESRO, ELDO and ESA are often described as the "space dimension" of a broader pattern of European integration and international cooperation (Sheehan, 2007, p. 86). Since its foundation, ESA successfully coordinated space activities at an international level thereby furthering integration of European space activities, by developing space programmes. ESA conducts research and development of space technologies and applications and develops and executes a European space programme and a European industrial policy (Rüttgers, 1989, p. 128; Kaltenecker, 1989, pp. 245–6). It is also tasked with the development and application of a long-term European space policy.

At the same time, the EU, has proven a motor for integration especially in questions of economic interest. The use of space technologies has developed into an increasingly important tool for the achievement of a considerable amount of EC/EU policies such as for instance transport infrastructure, environmental protection, and the information society. With two international organisations acting in the field of European space activities, questions concerning a closer cooperation between the two arose (Hobe et al., 2006, p. ix, 537). The EC's policy areas have increased considerably since its foundation in 1957. Given this development and the fact that the EC was a major potential user of space systems, it was only a matter of time that the EC would start considering space policy, a policy field to take within its ambits (Sheehan, 2007, p. 86).

Security concerns, differences in national strategies, size of investment, budget size and the principle of juste retour were among the reasons why European space collaboration was developed outside of the policy competence of the EU. However, external and internal changes after the mid-1980s—the change in the context of space and extension of policy competence of the EU in other areas such as research environment—seemed to have brought the EU a larger role in European space activities. Moreover, space would contribute as a symbolic feature of European integration in representing itself on the world stage. The commercialisation of space also called upon the EU to play an important role in European space collaboration. Thus, the third phase of space collaboration in Europe is characterised by the end of ESA's solitude as the only space-Europe, due to the EC's engagement in space policy, which resulted in an initial ad hoc collaboration with ESA.[23]

The third phase of the development of space activities in Europe thus deals with the first steps of the EC to enter the space arena.

3.3.1 The EC's Early Steps

For many observers of the development of the EC and today's EU to a space actor, the SEA of 1986 marks the starting point, as it added research and development to the Community's competencies, giving the EC the competence needed to get active in the field of space research and space technology development (Madders & Thiebaut, 1992; Reuter, 2007, p. 12). Formally the EU's (then the EEC's) involvement with space affairs began with the participation of the Commission as observer at the 1970 ESC (Hobe et al., 2006, p. 64; Madders & Thiebaut, 1992). This invitation was repeated for each subsequent ESC session, a tradition that has been extended to ESA Council meetings at ministerial level, where the Commission attends as an observer later (Madders & Thiebaut, 1992; Sheehan, 2007, p. 86; Peter, 2007, p. 100).

The idea that the European Community provided a logical framework within which all European integration efforts should take place had however first been suggested by the WEU in 1966, when the WEU Assembly called upon its member governments to "prepare for the inclusion of a permanent European space vehicle launcher development organisation with the framework of a future single European community." (Sheehan, 2007, p. 86; Madders, 1997, p. 121). The WEU suggested the European Communities should coordinate space policy and its execution (Madders, 1997, p. 121).

[23]This section is largely based on Suzuki (2003, p. 185).

The European Parliament (EP) was the first European institution to monitor events in the space field (Madders, 1997, p. 570), issuing two reports in 1979 and 1981, adding to the flow of reports and recommendations, coming from the Council of Europe and WEU's assemblies (Madders & Thiebaut, 1992). A third report followed in 1988. In its three reports, the Ripamonti Report (1979),[24] the Turcat Report (1981) and the Tokswig Report (1988), the EP requested the Commission to provide political recommendations for a EU space policy and to take the respective measures to implement it (Hutter, 1991, p. 156). While previously the EC had already engaged in R&D, it was only from 1988 that space aspects began to be treated in a consistent manner.[25]

The so-called Ripamonti Report and its subsequent EP "Resolution on Community participation in space research" are the first documents to call for a dialog with ESA, by asking the Commission to establish "relations with the ESA with a view to the coordination of space research programmes with Community projects" (European Parliament, 1979, p. 42). The emphasis was placed on assistance and cooperation. The EP

> urges the Commission, therefore, to *assist* the ESA in drawing up a comprehensive programme to meet the likely requirements of the Community Member States for the next 10 years, and to *cooperate fully* in ensuring that this programme is implemented as efficiently as possible (Ibid.) [emphasis added]

It was also the first Community document to highlight the impact of the participation of the Community in space research.

In 1981, the so-called EP Turcat Report urged the Council to call a European Space Conference at ministerial level. It emphasised "the need to promote the evolution of the role of the European Space Agency as the forum for cooperation in space policy" (European Parliament, 1981, p. 103).

Through the so-called Tokswig Report and its subsequent Resolution on European Space Policy, the EP called on the Commission to initiate the process to work out a coherent policy on space activities. It also proposed a "separate membership of the ESA for the European community, as a legal entity and requests the Commission and Council to prepare the appropriate negotiations" (European Parliament, 1987).

With the signing of the SEA, the EC possessed an interest in space for four reasons: as a focus for developing certain types of applied space technology, as a sponsor of research and development programmes, some of which have a space element; as a key force in shaping the market conditions in which the space industry operates; and as an increasingly active actor in the security field (Sheehan, 2007, p. 86). CNES published a report shortly after the SEA entered into force, concluding that its impact will probably be limited at least in the first years of 1990s in which ESA will remain predominant (Suzuki, 2003, p. 185).

[24]In 1979 the Committee on Energy and Research of the European Parliament and its Rapporteur Camillo Ripamonti issued a report entitled "Community participation in space research" cf. Hobe, Kunzmann & Reuter, 2006, p. 65.

[25]For a detailed description about the main events affecting R&D in general from 1971 consult Madders (1997, pp. 568–9).

Nonetheless by February 1990 it was agreed that a fresh look at ESA's industrial system was required, given the increasing economic integration at EC level. To do so, a working group was set up in March 1990 to study the post-SEA landscape. It started by identifying EC law sections, which were understood to be relevant to ESA and space activities. Finally, a conceptual comparison of EC economic law with ESA's industrial system was conducted. It was concluded that ESA and EC are fundamentally asymmetric. While the EC aimed at generating a European system by adjusting the juridical condition at the bottom of the economic system, leaving market forces to do their work, ESA is an active agent, forming industrial structure from the top downwards. Since both organisations are fundamentally unlike at conceptual level, the working group was satisfied that the postulate of complementary was well founded theoretically and conflicts between the two law systems were concluded to be unlikely (Madders & Thiebaut, 1992, pp. 130–1). At the same time, the ESA Council had introduced a working group to study the consequences of the SEA for ESA. As a result, the International Relations Advisory Committee (IRAC) presented a document entitled "Broad lines of cooperation between ESA and the EC Commission" on 7th June 1990 (Hobe et al., 2006, p. 68). This document highlighted the importance of ESA's independence in the cooperation with the EC (Ibid., p. 70).

In its 1988 communication, entitled "The European Community and Space: A coherent approach" (Commission of the European Communities, 1988), the Commission for the first time reacted to the EP request and recognised that space was more than a specific sector or activity. It also recommended closer collaboration with ESA, by acknowledging that space "is an area in which the Community now seems destined to play a broader and more active role" (p. 1). The Commission highlights that

> Europe is still neither active enough nor sufficiently well organised, when it comes to exploiting the applications of space technology, for which the ESA does not have the necessary resources at present (p. 8),

and

> is still without a cogent overall policy which incorporates technological, industrial, commercial, social and even defence aspects. (Ibid.)

Under the heading "a role for the community" the Communication sought to establish "the principle of a Community involvement in Europe's space effort, and defined the main orientations for a role, which would be complementary to ESA" (p. 7). The Commission highlighted that space is not an "isolated sector of activity or a specific technology" (p. 10) by virtue of its applications and consequently it has an impact on a growing number of Community competences and ambitions. By establishing the strength and weaknesses[26] of Europe's space efforts, the Commission

[26]"Europe's efforts remain in some respects selective: Europe holds a strong position in the areas it has targeted but it is less active and more dependent in others. Europe depends on the US for some operations in space and it has not yet embarked on the development of very heavy or recoverable launch vehicles, a factor which may restrict its autonomy of exploitation of space in the future: Europe is still neither active enough nor sufficiently well organised when it comes to exploiting the applications of space technology." (p. 8).

justified a "Community action" as "both possible and desirable" and referred to it as an "added value" (p. 11). It thus proposed for the Community to

> put the full weight of its democratic legitimacy and of its established role in European society behind the ambitious new programmes which the ESA has decided to launch, and to further enhance, by strengthening its links with the ESA, the political credibility of Europe's space effort on the international stage (p. 2)

and to

> frame policies that create favourable conditions for the development of the technologies needed for access to space and for optimum exploitation of space applications, on the one hand, and the integration of these applications in the socio-economic fabric of the Community and giving Europe's space industry a stronger presence in the world on the other (p. 11)

as well as

> to ensure, through organic consultation mechanisms, that the action by the Member States, the ESA and other European organisations (Eutelsat, Eumetsat, etc.) in the space sector (p. 11)

remain consistent with the action of the other organisations in this field and with Community law (p. 11). The Commission highlighted that

> the Community will naturally be mindful too of the need to ensure coherence and complementarity between its own activities and those of the ESA and other European organisations in this field. The Commission will pay particular attention to strengthening and developing its cooperation with the ESA on a more systematic basis and with mutual respect for each other's competences. (p. 4)

The role of the Commission was thus understood as a complementary one. The Communication further specified six action lines to guide the Commission's work in this area: (1) Research and technological development; (2) Telecommunications; (3) Earth observation; (4) Industrial development; (5) Legal environment and (6) Training. According to Madders and Thiebaut (1992) the Communication should be understood as an information and discussion paper, "focusing attention on a given field" (p. 122) and "containing seminal elements of a policy already being considered by the Commission but upon which no definite view has been reached" (Ibid.).

The Communication had two main consequences. The first was "to change space from a topic of passing interest into an organised area of policy consideration within the Community" (Ibid., p. 572), while the Community institutions still had to find their way in this area. The second one was the foreseen relationship with ESA, which now had to be elaborated as to reach progress (Ibid.).

3.3.2 *First ESA-EC Cooperation*

In 1989, the cooperation between the EC and ESA got formalised through the establishment of five joint committees, charged with exchanging information and perspectives on matters of common interest.[27] This development resulted from a meeting between the Commission President Delors and the Director General (DG) of ESA, Lüst, who had agreed on a "declaration of solidarity" concerning future cooperative relations, which would be instituted on the basis of respect for one another's competences (Madders & Thiebaut, 1992, Sheehan, 2007, p. 86).[28]

In 1991, the Commission founded an Advisory Panel under the lead of former ESA DG Gibson with the objective to prepare a report on the future role of the EC in the field of space. The subsequently published "Gibson Report" analysed institutional and political questions including those related to defence as well as ESA-EC relations (Suzuki, 2003, p. 186). It underlined the importance of space in many sectors of direct interest to the EC and indicated a growth of these (Commission of the European Communities, 1991, p. 34). It observed that the areas of EC interest in space-related affairs "are not totally covered" by the previously established ESA-EC Working Groups and proposed "[t]he joint steering group should itself address some of the horizontal questions rather than just monitor progress made in the working groups" (Ibid., p. 28). Additionally, it concluded "it might well be more effective if participation in each other's working groups or programme committees came about as a result of an understanding at the working level of the practical benefits." (Ibid., p. 28). One of the results of the report was that the Commission, as well as the European space community, began to understand the necessity for developing a coherent Earth observation policy in Europe in order to coordinate the Commission's agriculture and environment policies (Suzuki, 2003, p. 187).

The Gibson Report (1992) was the basis for the Commission's Communication entitled "The European Community and Space: Challenges, Opportunities and New Actions." (Commission of the European Communities, 1993) Therein the "Community interest" is justified by the impact of the EC's competences and policies on space activities as well as the impact of space on the implementation of Community policies (pp. 6–7). The Communication is thus proofing the Commission's ambition to become an actor in the space field (Suzuki, 2003, p. 187). At the time, the Commission was already using information derived from Earth Observation (EO) data in a variety of applications related "to its activities in key areas, such as land use, agricultural statistics, environmental research and monitoring" (Commission of the European Communities, 1993, p. 6), and development aid. As a consequence, the Commission understood itself as a potentially large customer for space-derived information. The Communication reiterates the reasons mentioned in the previous one for the Community to contribute towards Europe's space effort:

> the political, economic and industrial dimensions of the challenges and issues which need to be addressed collectively; the sectoral mission and competence of ESA, which is mainly an agency for the research and development of space technologies and systems; and the need

---

[27] i.e. Earth observation and the environment, telecommunications, research and technological development (R&TD), industrial competitiveness and the internal market, external/ international relations and commercial policy.

[28] The subsequent European Parliaments resolution (European Parliament, 1991b) deals with the international dimension of European space research and does not touch the question of cooperation with ESA.

> to integrate space activities into the broader socio-economic fabric of Europe (especially Earth observation and satellite communications which arc space-derived activities driven by objectives defined outside the space sector). (Ibid., p. 7)

The Commission went on in laying down the objectives and implementation of the community role:

> Firstly, it will act as a facilitator by helping to establish the appropriate conditions for the development of space applications markets and a competitive European space industry. Secondly, it will become an important customer of space-derived services (directly for its own needs and as a proxy for other users). This will reinforce the "demand-pull" influence on the development of space applications, especially for environmental research and monitoring. Thirdly, it will use its competences in the fields of trade and its commitment to international cooperation for the benefit of the space sector. (Ibid., p. 33)

The Commission proposed a symbolic programme called "Vegetation", co-funded with France, Belgium, Sweden and Italy for monitoring land use, dedicated about 100MECU to EO under several research areas as part of its fourth R&D programme and stepped up its level of cooperation by creating the Space Advisory Group (SAG), an ad hoc consultative body including representative from the Member States of the EC and ESA Executive (Suzuki, 2003, p. 187). Although "[i]t would have been more logical if the Commission would have entrusted ESA, as the typical European space entity, with the management of the project" (Suzuki, 2003, p. X), the Commission chose to work with CNES as Vegetation was strongly supported by France (Suzuki, 2003, p. 189).

The Space Advisory Group (SAG) had been established with the objective to centralise the informal practices of cooperation and the ad hoc cooperative mechanisms (Suzuki, 2003, p. 189). SAG met twice a year and provided an opportunity for high-level representatives of ESA, the Commission and relevant ministries of Member States to meet and discuss European space policy issues; thereby providing "an effective bridge between the ESA and EU decision—making process" (Ibid., pp. 189–90). It acted as the linchpin between the ESA Council and the EU Council of Ministers. Its activities as an advisory group were however limited to identifying areas were coordination was necessary. Its political influence was subject to the support from the Commission and Member States. SAG could have become an effective forum for discussions of European space policy if its statutes would have been upgraded to having an influence on the Commission. SAG also depended on the ideas of its Chairman. While during the chairmanship of Jean-Pierre Contzen (1992–1996), an enthusiast for using SAG as a political framework, its proposal for a common European policy for EO was successfully introduced in 1995 and adopted in both the ESA and EU Council. SAG lost its impetus to influence decisions at the level of the Commission under the chairmanship of Allgeier, who became the next chairman and who held a minimalist view on the Commission's role in space. SAG could have become the base for the European Space Council, which the EP repeatedly asked the Commission to consider (Ibid., pp. 189–91).

Also in 1992, Contzen was appointed as "space coordinator", tasked to facilitate inter-DG discussion and to formulate coherent space policy for the Commission. He was also responsible for organising the Space Coordination Group (SCG), a forum for representatives from the Commission's DGs. The SCG was supposed to provide a single view of the Commission's policy on space as to strengthen its position in negotiations with other entities. As with SAG, the SCG was limited to inter-DG coordination (Ibid., p. 191).

In 1991, the EP urged the Commission in its so-called "Rovsing Report" (European Parliament, 1991a) to define a comprehensive space policy. The report conceded ESA's competence "for the time being" (Madders, 1997, p. 574). Additionally, it noted among other things that while ESA was not politically equipped to "enforce an overall European space policy" (European Parliament, 1991a in Madders, 1997, p. 87), the EU in contrast possessed a huge range of legal and constitutional mechanisms designed to allow it to define and implement common policies. These mechanisms clearly gave the Community the opportunity "to play a crucial role in making it possible for Europe to reap the full benefits of ESA's outstanding achievements" (Sheehan, 2007, p. 87) and to "define broader and more complex goals, taking social political, economic, environmental and other considerations into account" (Ibid.).

Following the SEA, the Treaty on European Union (TEU), signed in Maastricht on 7 February 1992 (entered into force on 1st November 1993) further added to the competence of the EU in R&D. The contained provisions provide a basis "for pursuing a strategy of "synergy", "complementarity" and coordination" (Madders, 1997, p. 578) but are not supportive when it comes to dealing with another international organisation. The Maastricht Treaty can also be understood as the first step towards a political union, as it contains provisions on a Common Foreign and Security Policy (CFSP) (Ibid).[29]

The EP's Resolution entitled "Community and Space" of 6 May 1994 "calls upon the Commission to reinforce its coordination and cooperation with ESA" (European Parliament, 1994, p. 467) and to work together with ESA towards "the formation of a European Space Council" to ensure coordination and coherence within European space activities. The report inter alia refers to "external competitive threats" (Madders, 1997, p. 576).

In 1995, the EP organised the "European Space Forum", offering an opportunity for representatives from the Commission, ESA, industry and the user community to gather. It provided "almost for the first time" (Suzuki, 2003, p. 187) an opportunity to formulate a comprehensive European strategy beyond organisational and agency/industry boundaries. It offered a formal framework for the European strategy for space, something that the already existing business conferences could not offer. The issues discussed at the European Space Forum were directly reported to the SAG (Ibid.).

In view of the increasing importance of space activities, three Commissioners agreed to redefine and strengthen the SCG in 1996, equipping it with five permanent officials with the objective to assure a coherent and synergetic approach to space matters, to monitor Commission activities on space, to assure regular consultation

[29]For further information on Europe as an actor in space and security cf. Rathgeber and Remuss (2009), Remuss (2010, 2011b).

and dialogue with industry, national and international organisations and to coordinate socio-economic analyses (Ibid., p. 191). This "regeneration of the SCG" (Ibid.) increased the capability of the Commission to deal with space matters.

Under the framework of the SAG the Commission set up an ad hoc group of senior representatives from industry in January 1996. This "High Level Industry Working Group" presented specific recommendations to the Commission, thereby demonstrating the increasing importance of, and variety of opportunities for, the Commission in the space domain (Ibid., p. 187).

In 1996, the European Commission issued a Communication entitled "The European Union and space: Fostering applications, markets and industrial competitiveness" (Commission of the European Communities, 1996), which "confirms the terms of Parliament's resolution of 6th May 1994 referred to above. It also confirms the conclusions of the European Space Forum held on 6th and 7th November 1995" (European Parliament, 1998, p. 27) and laid down the roles of the "partners in the development of space technologies and applications in the European Union" (Commission of the European Communities, 1996, p. 10) as shown in the subsequent Fig. 3.1.

The Commission underlined that "ESA should remain one central piece of the institutional scheme, fostering European cooperation and strategic planning in the

THE PARTNERS IN THE DEVELOPMENT OF SPACE TECHNOLOGIES AND APPLICATIONS IN THE EUROPEAN UNION

1. **The European Community** : support to the development of the market, pilot and demonstration projects, methodological research...

2. **The governments of the European States and their space agencies or national programmes** : development of the space segment, exploratory and application missions, launching facilities...

3. **The European Space Agency** : development of the space segment, scientific, "Explorer" and "Earth watch" missions, launchers...

4. **The Space industry** : prime and sub-contractors for the space and ground segments, value added companies for services and products...

5. **The Space system operators** such as :

   - EUMETSAT for meteorology,
   - BUTELSAT for telecommunications,
   - EUROCONTROL for air traffic management

6. **The end users**

Fig. 3.1 Roles for European Actors in Space as laid down by the Commission in 1996 (Commission of the European Communities, 1996, p. 10)

space field" (p. 9). Moreover ESA "should reinforce the European technological base and remain a mainstay of European industry, through the design and implementation of space programmes for their specific objectives." (p. 9). The Commission for the first time also explicitly called for an improvement of

> structures of collaboration and cooperation to reinforce the synergy of the European space effort (…), maximise the return on investment made, and ensuring respect for the individual responsibilities of the various partners in developing a coherent overall strategy. (p. 27)

The relevant section deals mainly with the SAG and does not go into details about the roles of EU and ESA.

The EP in response to this Communication

> called on the Commission, when defining the guidelines for European space policy, to pay particular attention to the strengthening of the industry's technical and economic potential by means of measures promoting, on the one hand, innovative applications connected with manned space activities and with the major technological programmes and, on the other, technical and commercial applications which serve the mass markets (telecommunications) and major services (navigation). (European Parliament, 1998)

3.3.3 New Dimensions of Cooperation

According to Suzuki (2003), "one of the catalysts for the fusion of the ESA and the EU decision-making timetable was the question of European dependence on the U.S. satellite navigation system" (p. 193). European airlines as well as other users of satellite navigation services relied heavily on the U.S. Global Positioning System (GPS), which was completed by 1993. GPS was controlled by the military but provided signals with limited accuracy for civilian use as well. The U.S. Department of Defense (DoD) retained the option for discontinuing civilian usage in case national security was threatened, which was the case during the Kosovo crisis, resulting in disturbed air traffic in Greece and Italy. As a consequence, the European Commission's DG Transport initiated the idea of developing a European system as an alternative. While the EU lacked technical competence in space hardware development, it "was widely recognized that the ESA, as a R&D organization, alone would not be appropriate to plan and negotiate the program" (Ibid., p. 192). For the cooperatively developed programme to be successful, it was important "to have a common strategy and institutional linkage between ESA and the EU, something lacking throughout the history of European space collaboration" (Ibid.).

As in the previous phase of the development of European space collaboration, it was Britain, who provided for the breakthrough. Lord Sainsbury, British Minister for Space, was able to bridge Member States' space agencies and ministries in charge during the British EU Presidency during the first half of 1998. The result was the adoption of the exact same text by both Councils (ESA, 1998b). Sainsbury foresaw a clear division of labour. Accordingly, the EU was to provide the legal and regulatory framework for space and space-related activities, acting as customer, representing

Europe in international negotiations, while ESA was to provide the technical expertise (Suzuki, 2003, p. 192). The joint resolution can be seen as a "modus vivendi", being able to build on joint activities and sectoral joint strategy development already implemented in previous years, without providing for a binding framework for coordination or sustained joint activities between the two organisations (Cheli & Schrogl, 1999, p. 65). It is still a turning point providing for "a new era of cooperation between the two organisations" (ESA, 1998a, p. 1), demonstrating the willingness of the Member States of both organisations to strengthen the institutional leadership (Suzuki, 2003, p. 192). According to Hobe, Kunzmann, & Reuter (2006, p. 88) the two resolutions can be seen as a new impulse in the discussion on the cooperation between the two institutions. Accordingly, after a period of distrust since the entering of the EC in space policy, the two resolutions led to relaxation in the relationship between EU and ESA.

3.3.4 Turning to a New Chapter

ESA's Ministerial Council of March 1997 had called for a renewed mission for ESA as well as for an "evaluation of its specific role vis-à-vis other European and national institutions" (Cheli & Schrogl, 1999, p. 63). To do so, DG Rodotà created four expert study groups, advising him on (1) the overall evolution of ESA, (2) the rationalisation of resources in Europe, (3) the relationship with the EU and synergies with other sectors and on (4) the relationship with industry and harmonisation of technology. These groups met in 1997 and 1998 and their results were reflected in the first draft for a comprehensive European space strategy tabled in February 1998. In parallel, the ESA Council established a Council Working Group (CWG), tasked with the preparation of the next Ministerial Council Meeting foreseen for summer 1998. Given the resulting debate, it became clear that a Ministerial Council taking comprehensive decisions on the space policy issues was not feasible in 1998 (Ibid., pp. 63–4). It was instead scheduled for May 1999. The idea was largely driven by the experience of the Galileo programme. Rodotà's initiative was motivated by eight concerns of deficiencies in European space policy:

- low-public investments vis-à-vis the United States
- lack of understanding of space as an instrument and as an integral part of overall political, scientific, economic and military leadership
- U.S. dominance in innovative applications
- dependence on the U.S. for satellite navigation
- division of strategy and actions at European, multinational and national level
- limited dual-use synergy
- lack of ESA commitment to space applications
- duplication of facilities and operations in ESA and national operations.

To overcome these deficiencies Rodotà called for "More and Better Europe", meaning a more common and coordinated approach among all the actors in Europe.[30]

As a result of the joint reflection on practical measures to implement closer co-operation, a Commission Working Document of 1999 on developing a coherent European space strategy draws preliminary conclusions of the joint report on synergy between the Community and ESA (Commission of the European Communities, 1999, p. 9). It remains limited to practical measures improving consultation and coordination and while not going into details with regard to the roles of each actor.

Based on this principle the ESA Ministerial Council in Brussels issued a Resolution entitled "Shaping the Future of Europe in Space", asking the ESA Executive to prepare a fully developed strategy together with the European Commission to be delivered by the end of 2000. The Council of Ministers of the EU followed this example, asking the Commission to "prepare, as a joint exercise with the ESA and using a single and common consultative framework with all interested parties, a comprehensive document on a coherent European strategy for space" (Suzuki, 2003, pp. 197–8; Council of the European Union, 1999). These two resolutions again provide a turning point to the degree to which the Commission became involved in space. It also granted significant importance to the Commission as the central agency for European strategy development (Suzuki, 2003, p. 198). Additionally, it was practically the first time for both organisations to develop together a common strategy, forming a single team through the newly created Joint Space Strategy Advisory Group (JSSAG) consulting with Member States and industry. During 2000 this joint team consulted on four subjects with the objective of preparing a strategic document: space-based information and communication services; science; research and space exploration; Common Foreign Security Policy (CFSP)-consideration (cf. Rathgeber & Remuss, 2009; Remuss, 2010, 2011b); and access to space. The WEU expertise was also consulted. The result was a preparatory strategic document, adopted by both the EU Council of Ministers and the ESA Ministerial Council in September 2000 entitled "Europe and Space: Turning to a new chapter" (Commission of the European Communities, 2000; ESA, 2000). The documents refer to the strategy as proposed as "a further milestone in establishing a coherent approach for space in Europe" (Ibid.). The Commission is envisaged to take an "active role in the implementation of the strategy" (Ibid.). ESA is said to have become "Europe's Space Agency in a wider sense" (Ibid.). ESA and the Commission are said to be "ready to enter a new operational relationship with ESA to work together under their respective competencies for the implementation of this strategy" (Commission of the European Communities, 2000, p. 16). Moreover, the transferral of the WEU Satellite Centre to the EU is foreseen. The strategy calls for "formal links between the Union and ESA, which would allow a straightforward practical implementation of common activities" (Ibid.) as "a more sustainable path should be established for implementing a common European strategy" (Ibid.). Acknowledging the importance of preserving the flexibility and know-how of ESA in space activities, the document calls upon ESA to assume further responsibilities in the context of the two flagship programmes Galileo and

[30] This section is based on Suzuki (2003, p. 197).

Copernicus. In addition to acting as a space agency for its Member States, ESA is asked to become the "implementing agency for the development and procurement of the space and ground segments" (Ibid.) associated with Galileo and Copernicus.

The above-mentioned SCG was reorganised following the Commission's drafting of the European Strategy for Space as the space policy unit, strengthening the Commission's policy planning function and communication with industry and ESA (Suzuki, 2003, p. 191).

3.3.5 *The Three-Wise-Men-Report of 2000*

After the founding of the EU with the Maastricht Treaty, the EU began to be envisaged as the major political and decision-making instrument for a "European space policy" with ESA being understood as the implementing agency. This was also stressed by the 2000 Three-Wise-Men-Report (Marchisio, 2006, p. 197).

Rodotà called a group of high-level executives—Bildt (former Prime Minister of Sweden), Peyrelevade (President of Credit Lyonnais) and Späth (CEO of Jenoptik AG) to provide independent advice for promoting the evolution of the ESA-EU relationship and the necessary institutional change (Suzuki, 2003, p. 199). The resulting so-called Three-Wise-Men-Report of 2000 (Bildt, Peyrelevade, & Späth, 2000), called for the establishment of a coherent ESP, incl. the strategic dimension of space and the demand for space applications derived from other policy fields of the EU (Hobe et al., 2006, p. 540), and was thus considered a turning point of the EC's engagement in the field of space policy (Kunzmann & Reuter, 2004). Despite the call for a European space policy, especially by the European Parliament the Commission did not work on an overarching policy but limited itself to the coherent integration of space applications in the existing sectoral policy areas and the deepening of the relationship with ESA. It did not follow a holistic approach but focussed on the sectoral benefit of space technology to community policies (Reuter, 2007, pp. 12–3).

The three Wise Men were convinced that space policy needed to enter a new phase, becoming an "integrated aspect of the overall efforts of the countries and institutions of the European Union" (Bildt, Peyrelevade, & Späth, 2000, p. 2) and "an integral part of the efforts of European integration to enlarge, to deepen and to allow Europe to play a more important role in the world" (Ibid.). They thus "want policies to be set more clearly, commitments to be stronger, institutions to move closer to each other and artificial barriers and blockages of the past to be taken away." (Ibid.) About the institutional set up, the Wise Men saw "a need for changes, especially in the relationship between ESA and the EU." (Ibid., p. 7) Seeing space no longer as an exclusive and separate activity, they recommended a "closer institutional integration, thus ensuring the place of space issues in the overall evolution of European policies." (Ibid.) Considering the "need for the institutions starting to work together now" as well as "the need for a process of institutional convergence", they do not explicitly "exclude bringing the present ESA within the treaty framework of the European

Union" (Ibid.). They explicitly "urge[d] the institutions to initiate a process of institutional convergence" (Ibid.) and suggested the following institutional set-up and division of labour:

- European Council: to define the European Space Policy and the guidelines for its implementation (could be reviewed every five years),
- ESA: to be the space agency of Europe setting and implementing co-operative programmes (on a reinforced cooperation basis allowing the full participation EU Member States, the European Commission and other countries of Europe)
- European Commission: to define the regulatory framework under which space activities are conducted, representing Europe in world-wide fora allocating frequencies and defining market rules, bringing together user interests around common objectives, the European Commission should also be a contributor to ESA programmes and, as such, a member of the ESA Council.
- EP: to be given the opportunity to regularly discus and review the European Space Policy. (Ibid., p. 8)
- ESA and the European Commission: to jointly develop the European Space Policy and propose it to the European Council and the European Parliament (Ibid., p. 8).

While suggesting ESA should become the de facto space agency of the EU, the three Wise Men explicitly underlined that the instrument of reinforced cooperation should not be restricted to only the present members of the EU but be open to all European countries (Ibid., p. 11). Considering the growing importance of space systems for society and markets world-wide, the three Wise Men saw a need for the European answer to consist of a more integrated approach, i.e. space activities should be understood as part of the wider political and economic strategy. They considered the complementary decisions of the ESA Council and the EU Council to develop a joint European Strategy for Space by the end of 2000 and the preceding joint ESA-European Commission document on a European Strategy for Space (ESS) (Ibid., p. 5).

Shortly thereafter, the Council set up the European Union Satellite Centre (SatCen; previously known as EUSC[31]) within the EU in 2001 through a Council Joint Action (Council of the European Union, 2001). This was done by incorporating the WEU Satellite Centre of 1991 and its existing features and structures. SatCen became operational in 2002. It is in charge of supporting the CFSP and the European Security and Defence Policy (ESDP)[32] through space-based intelligence (Marta & Stephenson, 2016, p. 104).

[31]Cf. Davara (2000).

[32]For more information on the role of space in the CFSP cf. Rathgeber and Remuss (2009), Remuss (2010, 2011b).

3.3.6 A European Space Strategy

On 16 November 2000 both the Council of Ministers of the EU (cf. Council of the European Union, 2000) and ESA's Council (cf. ESA, 2000) adopted a European Space Strategy based on the Joint Resolution as well as the Three-Wise-Men-Report.[33] The impetus for this process was derived from a position paper produced by the French space agency CNES and the German space agency DLR in cooperation with the Italian space agency ASI of March 1998 (Dupas, Janischewski, von Kries, & Schrogl, 2001, p. 105).

The strategy structures Europe's future efforts around three objectives:

- "strengthening the foundation for space activities thereby preserving independent access to space and ensuring a broad technology base
- enhancing scientific knowledge for a better understanding of our planet, the solar system and the universe
- reaping the benefits for markets and society through a demand-driven exploitation of our space capabilities with an open eye for applications of space-based information for the benefit of the Common European Security and Defence Policy" (Council of the European Union, 2000, p. 1).

Previously, the Commission had in a Communication (cf. Commission of the European Communities, 2000) foreseen the following task-sharing. The Commission was "to take an active role in the implementation of the strategy" (Ibid., p. 3) by establishing "the right political and regulatory conditions for space activities, in line with and in support of Community policies." (Ibid.) "Building on its achievements and its technical expertise" (Ibid., p. 4), ESA was understood to remain the principal programming and funding agency.

The documents envisaged for the Commission and the ESA Executive to set up a Joint Task Force (JTF) to "stimulate the reflection on a possible unified setting for Member States to review the strategy and its implementation on a regular basis, in order to propose permanent arrangements before the end of next year" (Bildt, Peyrelevade, & Späth, 2000, p. 17) and to "elaborate proposals for framework arrangements for the management of joint projects and outline a coherent approach with respect to candidate countries" (Ibid.).

The JTF delivered its first report in December 2001, recommending closer cooperation between ESA and the EU and proposing to set up a "Space Council" (comprising the EU Council and the ESA Ministerial Council) in order to develop a European space policy. This was to be presented to the European Council, where heads of State and governments can consider space policy in full, including its security and defence aspects (Peter, 2007, p. 101; Suzuki, 2003, p. 200). Accordingly, while "ESA has a record of technical achievements and know-how" (Commission of the European Communities, 2001, p. 5), space issues need to acquire "full political legitimacy" (Commission of the European Communities, 2001, p. 5), requiring "a closer co-

[33] The joint document has been released as an annex to the Three-Wise-Men-Report.

operation of ESA with the EU" (Ibid.), as "[s]pace has entered a new phase" (Ibid., p. 30).

> It has evolved from being a research endeavour and instrument to become a unique and critical technology enabling Europe to address and achieve a large number of the goals embedded in policies such as economic growth, information society, transport and energy infrastructure, environment protection and peace-keeping. (Ibid.)

Thus, "[s]pace must become an integral component of the EU's core policies" (Ibid., p. 30). The JTF among other things recommended "perform[ing] a thorough analysis to define ESA's future additional roles as the "implementing agency" of the EU." (Ibid., p. 32).

Through the joint preparation of the strategy document, the Commission and ESA are said to have entered a new relationship, which is to be built up further, through a fruitful interaction between administrations at national, intergovernmental and Community level (Busquin, 2001). Bringing ESA and EC together in a joint exercise is a tremendous progress in itself (Dupas, Janischewski, von Kries, & Schrogl, 2001, p. 105).

In the same year on 13 November 2001, the Council issued its first Copernicus-dedicated resolution (Brachet, 2004, p. 12).

3.3.7 Green and White Paper on European Space Policy

In 2002 and 2003 a series of workshops on European Space Policy was initiated after a request by the EP (Köpping Athanasopolous, 2016, p. 89), as to better determine and prioritise space policy options for Europe. The first European Space Policy Workshop on "Fundamental Issues" took place in Leuven in September 2002, and the second to follow at the beginning of 2003 dealing with the stakeholders and their interests.[34] Shortly thereafter a public consultation about Europe's place in outer space affairs was initiated, aiming at formulating a Green Paper on European Space Policy as a cooperative attempt by the EU's Commission and ESA (Madders & Wouters, 2004; Hobe, 2004, p. 25). It was seen as a first step towards a White Paper. Europe's strength and weaknesses in the space sector were assessed "to launch a debate on its space policy involving all players including national and international organisations, the European space industry and its users and Europe's scientific community and citizens" (Peter, 2007, p. 101). During the consultation phase on the Green Paper ten workshops were organised by the EC-ESA JTF or by Member States. Each took place in a different EU Member State's capital and each was themed in order to target different sectors of the space community (Commission of the European Communities, 2007b, p. 7).[35] Thus, the Green Paper process was instrumental in

[34]Further workshops were suspended given the pending outcome of the Green paper consultation process. Thus, the third workshop on human spaceflight took place on 8 September 2003. Cf. Madders and Wouters, (2004).

[35]i.e. industry, science, commercial applications, government etc.

bringing together stakeholders, observers and analysts around the issue of how to increase the relevance of space in Europe (Mazurelle, Wouters, & Thiebaut, 2009, pp. 18–19).

The EP also tabled a report (cf. European Parliament, 2003) on the Green Paper in 2003. Therein among other things the EP "[r]eaffirms its proposal, already submitted a number of times, for a European Space Conference to be convened in 2005 in order to widen and deepen the ways in which Europe can operate in this area" (European Parliament, 2003, p. 7).

Based on the Green Paper process, the EC released its White Paper entitled "Space: A New European Frontier for an Expanding Union—An Action Plan for Implementing the European Space Policy" (cf. Commission of the European Communities, 2003). Highlighting that space technology, infrastructure and services were an essential support to both the CFSP and ESDP. The 2003 White Paper confirmed the general political and strategic importance of space:

> Europe needs an extended space policy, driven by demand, able to exploit the special benefits space technologies can deliver in support of the EU's policies and objectives: faster economic growth, job creation and industrial competitiveness, enlargement and cohesion, sustainable development and security and defence. (Commission of the European Communities, 2003, p. 4)

It included an action plan for European cooperation in space with concrete proposals for the allocation of funding for European space programmes. Accordingly, a European space policy could only be achieved by concentrating efforts within the European framework. Among the challenges the White Paper mentioned

> to identify and encourage the most effective sharing of tasks and responsibilities between the EU, ESA, Member States, national agencies and industry for determining and continuously delivering the benefits of space to the Union and its citizens. (Ibid., p. 37)

It highlighted that

> successful implementation requires a system of governance in which the roles of the players are clearly understood, the tools for careful coordination properly provided and processes of accountability put in place. (Ibid.)

Concerning the institutional setting it recommended the positioning of ESA within the EU framework after an eventual entry into force of the constitutional treaty (Hobe et al., 2006, p. 540).[36] The White Paper emphasised the ability of space applications "to satisfy the needs of the citizens and to respond to the Union's political objectives" (Commission of the European Communities, 2003, p. 5) and therefore called for statute changes to give the EU "new responsibilities for driving, funding and coordinating activities within an extended space policy" (Ibid.). This had major implications both for the direction in which the Commission was seeking to take the Union, and for the relationship between the EU and ESA (Sheehan, 2007, pp. 87–8). Accordingly, the "European Space Agency should take the lead in elaborating, proposing

[36] "As a key component of the European space policy, ESA must eventually acquire legitimacy in the EU framework." Cf. Commission of the European Communities, 2003, p. 36.

and developing the required solutions" (Commission of the European Communities, 2003, p. 39). It is underlined that space and its international cooperation "should also be in function of serving the widest-possible spectrum of the EU's objectives" (Ibid., p. 21). It recommends that "the European Commission should develop, benefiting from its collaboration with ESA, a strategy for international space co-operation for the next decade with particular emphasis on supporting the EU's CFSP" (Commission of the European Communities, 2003, p. 21), underscoring the EU's increasing willingness to link space activities with more global foreign policy objective (Peter, 2007, p. 103).

According to the White Paper the implementation

> of a European Space Policy would come in two phases: the first (2004–2007) will consist of implementing the activities covered by the recently agreed FA between the European Community and ESA; and the second (2007 onwards) will start after the coming into force of the European Constitutional Treaty. (Commission of the European Communities, 2003, p. 4)

During the first phase the two actors will be able to set common objectives, while during the second phase "ESA should (…) be positioned within the EU framework and its Convention modified accordingly" (Ibid., p. 38). Observers refer to this suggestion as "ambitious governance revamping" (Mazurelle, Wouters, & Thiebaut, 2009, pp. 19–21) and interpret the division in two phases as a hint about ESA being foreseen to become integrated in the EU after the coming into force of the Constitutional treaty (Reuter, 2007, p. 22).

At the same time, ESA's DG Dordain issued a strategy paper entitled "Agenda 2007" (cf. Battrick, 2003), which laid down his plans for his term of office. Therein he highlighted the need for ESA's position to be well defined within the framework of the EU institutions as to avoid ongoing justification concerning its role (Ibid., p. 16). He saw "an urgent need to prepare for the new era, by establishing a relationship of trust between ESA and the European institutions" (Ibid.). Similar to the Three-Wise-Men-Report, he proposed a possible division of labour as the EU clarifying the requirements and ESA being responsible for the technical execution.

3.3.8 Draft Treaty Establishing a Constitution for Europe

With the SEA introducing R&D as one of the Community's policy areas, and the Treaty of Maastricht, establishing a CFSP, the next step in this development was the inclusion of a specific competence for the EU in space matters through the Draft Constitutions of the Brussels Convention (Kunzmann & Reuter, 2004). A constitution for Europe was sought, given the increasing frequency of treaty revisions and the resulting need for greater transparency and simplification (Bainbridge, 2003, p. 87). The first step towards a constitution had been taken by the Laeken Intergovernmental Conference (IGC) in 2001. A European Convention was formed shortly thereafter with the objective to draft the Treaty establishing a Constitution for Europe (TCE).

The final draft (cf. The European Convention Secretariat, 2003) was agreed upon and submitted to the President of the European Council during the Italian Presidency in 2003. It contained provisions (Art. III-13 para 3) on a "special category of competences shared between the Union and Member States" (Kunzmann & Reuter, 2004, p. 60):

> In the areas of research, technological development and space, the Union shall have competence to carry out actions, in particular to define and implement programmes; however, the exercise of that competence may not result in Member States being prevented from exercising theirs. (Art. III-13 para 3)

Art. III-155 contained more detailed provisions on how the EU is to exercise its competences (Kunzmann & Reuter, 2004). Accordingly,

> "1. [t]o promote scientific and technical progress, industrial competitiveness and the implementation of its policies, the Union shall draw up a European space policy. To this end, it may promote joint initiatives, support research and technological development and coordinate the efforts needed for the exploration and exploitation of space.
>
> 2. to contribute to attaining the objectives referred to in paragraph 1, European laws or framework laws shall establish the necessary measures, which may take the form of a European space programme." (Article III-155)

The inclusion of space in the new Constitution reflects

> the growing recognition among EU leaders that space systems and related technologies can be used to provide beneficial products and services to European citizens, while delivering a powerful competitive edge to European industry in the global market place. (Times Higher Education, 2004)

Former French President d'Estaing appreciated the Constitution especially regarding the allocation of competencies among the Union and Member States by making it clear "who does what in Europe" (Ibid.). The "unprecedented references to space in a European Treaty" was considered "a true milestone, laying the way for a new era in Europe's history in space" (Ibid.) by the community. The previously mentioned Green Paper had already laid down the reasons for including space in the Constitutional Treaty.

> First, space is (…) extraterritorial and (…) calls for human resources and financial support that go beyond purely national constraints. (…) space presents unique opportunities for the development of a broad range of highly beneficial civil and public services, including environmental monitoring, navigation and observation services as well as defence- and security-related services. A pan-European institutional framework for space activities would help to focus and rationalise Europe's space activities, providing a common vision and a common set of rules. (Commission of the European Communities, 2003b, p. 2)

Even though the treaty had been signed on 29th October 2004 by all 25 Member States, it could never be ratified due to the rejection of the text in referenda in France and The Netherlands. Thus, the second phase of the European Space Policy as foreseen[37] already by the White Paper in 2003 could not be started (Hesse & Hornung, 2012, p. 187). A similar text was later included in the Lisbon Treaty however, the treaty that was adopted as a consequence of the rejection of the Draft Constitution. Before that, Barroso, the new EU Commission President had transferred space in 2004 from the DG for Research to the DG on Industry and Enterprise, thereby emphasizing that Europe's engagement in space goes beyond research, having an important industrial dimension and being linked to security[38] (Verheugen, 2005, p. 93).

3.4 Phase 4: From Framework Agreement to the Lisbon Treaty—the Era of the Space Councils (2004–2009)

The fourth phase of the development of space collaboration in Europe has been characterised by the formalisation of the EU-ESA cooperation through the ESA-EU Framework Agreement (FA) (cf. Council of the European Union, 2003). At the same time, space applications became increasingly used in the area of security and environment. The Commission's continuous interest in space technologies resulted in space as an area of R&D, being first incorporated into the Draft Constitution and finally the Lisbon Treaty. The period from the FA to the Lisbon Treaty can also be referred to as the era of the Space Councils.

3.4.1 The ESA-EU Framework Agreement: Formalised Cooperation

In October 2003 ESA and EC agreed on the FA, which was extended twice since (2007 and 2012) (European Parliament, 2014, p. 48). The FA has two main objectives: first the coherent and progressive development of an overall European space policy; and secondly

> the establishment of a framework providing a common basis and appropriate operational arrangements for efficient and mutually beneficial cooperation between ESA and the Commission, fully respecting their institutional settings and operational frameworks. (Council of the European Union, 2003, p. 5)

---

[37] According to the White Paper the implementation "of a European Space Policy would come in two phases: the first (2004–2007) will consist of implementing the activities covered by the recently agreed FA between the European Community and ESA; and the second (2007 onwards) will start after the coming into force of the European Constitutional Treaty".

[38] For further information on Europe as an actor in space and security cf. Rathgeber and Remuss (2009), Remuss (2010, 2011b).

Institutionally, the FA replaced the previous ad hoc structure for cooperation.[39]

Hobe et al. (2006) understand the FA as "an important first step, formalising the relationship between ESA and EC and committing the two organisations to cooperate." (p. x). The FA laid down the principles of cooperation (Art. 2), the fields of cooperation (Art. 3) and the joint initiatives (Art. 4) of EC and ESA. The Preamble mainly shows the aim and wider importance of space as a policy field and explicitly highlights the maintenance of the "respective distinct tasks and responsibilities" of EU and ESA. Article 1 clarifies "the coherent and progressive development of an overall European Space Policy" and

> the establishment of a framework providing a common basis and appropriate operational arrangements for an efficient and mutually beneficial cooperation between the Parties with regard to space activities in accordance with their respective tasks and responsibilities and fully respecting their institutional settings and operational frameworks. (Council of the European Union, 2003, p. 5)

regarding the aim of the FA. This policy

> shall link the demand for services and applications using space systems in support of the Community policies with the supply of space systems and infrastructures necessary to respond to that demands. (Ibid.)

According to the FA, the cooperation of EU and ESA aims at bringing space

> to a wider political, economic, scientific, environmental, and social framework more directly to the service of European citizens. The closer cooperation between ESA and EC shall be based upon the specific complementary and mutually reinforcing strengths of the two organisations in an efficient and mutually beneficial manner in order to avoid any unnecessary duplication of effort. (Ibid., p. 3)

The FA distinguishes between five forms for joint activities (Art. 5) "without being limited to" them:

(1) "the management by the ESA of European Community space-related activities in accordance with the rules of the European Community;
(2) the participation by the European Community in an optional programme of the European Space Agency, in accordance with Article V.I.b of the ESA Convention;
(3) the carrying out of activities which are coordinated, implemented and funded by both Parties;
(4) the creation by the Parties of bodies charged with pursuing initiatives complementary to research and development activities, such as the provision of services, the promotion of operators formation and the management of infrastructures;
(5) the carrying out of studies, the organisation of scientific seminars, conferences, symposia and workshops, the training of scientists and technical experts, the exchange or sharing of equipment and materials, the access to facilities, and the support of visits and exchanges of scientists, engineers or other specialists".

---

[39]i.e. the ESA–EC JTF and the Joint Space Strategy Advisory Group) with the introduction of the Space Council.

Thus, while it does not dispense with the need to conclude specific agreements for joint projects, the FA seems to deal with all aspects involved in the cooperation of EU and ESA (Smith & Hörl, 2007). It lists possible cooperation models, which have also existed prior to the agreement. But it does not contain guidelines concerning their implementation and thus does not go beyond the reproduction of the status quo (Hobe et al., 2006, p. 544). Additionally, no solution had been included for the diverging financial mechanisms (Reuter, 2007, p. 21). While considering the FA a "milestone", Smith and Hörl (2007) understand it as "only a first step in the real institutional convergence needed between ESA and the EU" (p. 220). The FA provides general guidelines for the relationship between the EC and ESA, but equilibrium for an efficient, cost-effective and political sustainable alliance between the institutions still has to be found (Smith & Hörl, 2007). The FA formalises the status quo as this was the only compromise possible (Hobe et al., 2006, p. x, 544–5).

One of the advantages of ESA is its funding through the principle of "geo return", "juste retour", "fair return" or "return coefficient". ESA Member States fund ESA's activities according to their own level of interest in the organisation's activities (Gibbons, 1986, p. 30). Within the principle of geographical distribution (Art. VII para. 1 lit. c ESA-C and complemented by the preference clause set out in the same provisions). ESA provides a guarantee for Member States that their industry will receive contracts out of the amount ESA spends on contracts at a value linked to the contributions made by the respective Member States to ESA's mandatory and optional programmes. The principle of geographical distribution thus does not result in a financial but rather in an industrial return (Hobe et al., 2006, p. 557). While a coefficient of one could so far not be achieved for all Member States, "there is a trend towards a levelling off of the disparities in return coefficients among Member States" (Gibbons, 1986, p. 30). Hobe et al. (2006) claim that

> [t]he principle of geographical distribution contributes to ESA's success and to European space activities to a considerable extend; without the existence of such fair return the accomplishment of European space activities would hardly be conceivable. (p. 558)

Reuter (2007, p. 10) too refers to the principle of geo-return as the "formula for success". Its abolishment would shake ESA's very foundations. There is no principle of geographical or fair return applicable to the EC budget. The main element of the EC's economic policy is the internal market with its four fundamental freedoms (free movement of goods, services, capital, and persons) and the Community's objective to promote and preserve the maximum level of competition community-wide. The EU's financial order and its economic and industrial policy are governed by the postulate of maximum competition (Hobe et al., 2006, pp. 558–9). Some authors even argue that the principle of geo return would be contrary to the rules of the Common Market and the competition policy of the EU (Reuter, 2007, p. 10).

The main novelty, as introduced by the FA, is the introduction of the Space Council, which is foreseen by Art. 8. The objectives of the Space Council are identifying actions required; making recommendations, advising the parties on ways to enhance cooperation consistent with the principles set out in the FA reviewing the effective and efficient functioning of the FA. The Space Council is not equipped with decision making power, legally binding to both ESA and the EC and as such does not "constitute a new, superior organ" (Hobe et al., 2006, p. 544). Instead the Space Council can be understood as a meeting of both the European Council and the ESA Council at ministerial level at the same time. Thus, the establishment of the Space Council does not render the need for separate decisions in both organisations unnecessary. It does however allow for a more coherent coordination of the respective decision-making procedures, without however simplifying them (Ibid., p. x, 544–5). The procedures of the Space Council are quite complex in nature and thus do not allow for a flexible decision-making mechanism on a short-notice (Reuter, 2007, p. 21). Hesse and Hornung (2012) thus propose a two-fold assessment of the FA. While on the one hand formalising the existing cooperative mechanism providing for a permanent communication platform, the Space Council on the other hand lacks any decision-making powers and is thus limited in its evolution.

3.4.2 *The First Space Councils: Trying to Define Roles*

The first European Space Council was held in Brussels on 25 November 2004 and involved all EU and ESA Member States. The objective was "to discuss major policy issues concerning the definition of a European Space Programme" (cf. Council of the European Union, 2004) based on the questionnaire drawn up by EU and ESA. The resulting "First orientations on the preparation of the European Space Programme" do not mention the ESA-EU relationship. This is particularly striking as the questionnaire had included a section on "governance" calling for a new equilibrium "making best use of existing capacities and strengths and avoiding duplication of tasks" (Ibid., p. 3).

The Commission responded to the orientations of the first Space Council by publishing a Communication entitled "Preliminary Elements of the European Space Policy" (cf. Commission of the European Communities, 2005), produced by the EC-ESA joint secretariat. The document has been discussed with the Member States in the High Level Space Policy Group. The joint secretariat aimed to further develop these elements based on the orientations of the second Space Council scheduled for 7th June 2005 as to be able to develop a more concrete proposal for a European Space Policy and related Programme to be submitted to the third Space Council (Ibid., p. 3). In a section entitle "roles and responsibilities", the paper highlights the objective of achieving a "clear and complementary task allocation and in accordance with existing legal provisions." (Ibid., p. 6) Accordingly the role of the EU should be:

- "to define the priorities and requirements for space-based systems at the service of the EU's main objectives and policies and citizens' needs;
- to aggregate the political will as well as user demand in support of these;
- to ensure the availability and continuity of services supporting EU policies by funding

- relevant up-stream research activities; purchasing services or securing the deployment and
- operational phases of space systems, as appropriate; and in due course stimulating user funding;
- to ensure integration of space-based systems with related ground and in situ systems in
- promoting the development of user-driven application services supporting EU policies;
- to create an optimum regulatory environment to facilitate innovation;
- to promote coordination of the European position in international cooperation." (Ibid.)

For ESA the document foresaw the following role:

- "to support the technical specification of the space segment of space application programmes, taking particular account of EU requirements;
- to develop and implement space technologies, in particular in access to space, science and exploration;
- to pursue excellence in scientific research in, of and from space;
- to advise the EU on space segment requirements to support availability and continuity of services;
- to implement international cooperation within the ambit of ESA-led programmes." (Ibid.).

The document separates "roles and responsibilities" from "governance". It held that "[t]he closer integration between space activities and the policies of the EU" (Ibid., p. 7) the greater the "effectiveness of both those policies and of space policy" (Ibid.). Given the need for complementarity of the space-related actors' activities, Europe would need to optimise its governance scenario for space (Ibid.). Accordingly, "ESA will coordinate efforts to restructure space-related public infrastructure, with a view to increasing the efficiency of the programmes" (Ibid.).

The second Space Council, held in June 2005, foresaw a general exchange of views on the further definition of European Space Policy and the European Space Programme (ESP). Its objective was "to give clear political signal on the second set of orientations … relating to the future development of ESP" (cf. Council of the European Union, 2005a). The resulting orientations, which were understood to form the basis of "fully elaborated draft of a European Space Policy" (Council of the European Union, 2005a, p. 4), dealt in detail with strategy, roles, and responsibilities, priorities within the ESP and key principles of implementation. Among the "roles and responsibilities" ESA was for the first time understood as a development agency: "ESA and its Member and Co-operating States will develop space technologies and systems, supporting innovation and global competitiveness and preparing for the future" (Ibid.), while the EU was given the role of the "lead in identifying and bringing together user needs and to aggregate the political will in support of these and of wider policy objectives" (Ibid.). ESA's role was thus downgraded to "effective coordination with ESA of the European position in international fora" (Ibid.). ESA's main tasks together with the Member States were the "exploration of space and on the basic tools on which exploitation and exploration of space depend: access to space, scientific knowledge and space technologies" (Ibid.) as well as on "a voluntary basis" the support of "the technological preparation, including validation, of space systems responding to user needs, including those relevant to EU policies" (Ibid.).

The third Space Council, which took place on 17 November 2005, mainly dealt with the draft orientations on Copernicus following the meeting of the High Level Space Policy Group and a Presidency discussion paper entitled "First Reflections on International Relations". Thus, in terms of roles and responsibilities of the EU and ESA, it mainly confirmed the ones identified in the orientations of the second Space Council and added that Copernicus is to be understood as an EU-led initiative (Council of the European Union, 2005b). Reiterating the conclusions of the second Space Council, which laid down that Galileo and Copernicus are EU flagship programmes, the third Space Council among the "principles to be respected in the representation of Europe versus Third Parties" allocated the lead "to the overall representation of Europe regarding GALILEO and Copernicus, after consultation of Member States and duly taking into account the interests of ESA" (Ibid., p. 7) to the Commission. "The Commission shall also take the lead with a view to any market regulating measures to be introduced internationally." (Ibid.)

> Following the same line of argument, ESA shall take the lead as to the overall representation of Europe in the areas of science, launchers, technology and human space flight, after consultation of Member States and duly taking into account the interests of the EU. (Ibid.)

Building upon Agenda 2007 released in 2003, ESA's Director General published a "framework of action for achieving the wide-ranging objectives of Member States and for adapting ESA to a changing environment" as "Agenda 2011" (ESA, 2007a) in 2007. While ESA is recognised as a globally important agency in its core activities of science and exploration, human spaceflight and partnership in the ISS, and launchers, the future objective is to develop "beyond this, to make ESA a model for underpinning the use of space in the world today and specifically in the context of Europe's growing needs." (Ibid., p. 8) ESA will shift its focus on promoting space applications in areas such as security, disaster relief or maritime surveillance,[40] thereby adopting a user-centred approach. It aims to take the role of "promoter of the space element in an overall system" (Ibid., p. 9).

3.4.3 *The Commission's Impact Assessment: Four Policy Options*

Since February 2007, the Commission's Impact Assessment of the European Space Policy (Commission of the European Communities, 2007b) has been released.

[40] For more information on the use of space applications in maritime surveillance cf. Remuss (2010b, c, 2011c, d, 2014).

Therein problems of its implementation have been highlighted. Accordingly, given the innovative character of the sector, requiring long-term technology investment, the private sector cannot alone carry both the technological and financial risks (Froehlich, 2008, p. 68). Additionally, space policy is understood as being of strategic interest to many States, resulting in a rise of the number of space faring nations (Ibid.).

It analysed four policy options to achieve the proposed objectives: (1) no European Space Policy, (2) increased coordination and growing use of space applications to deliver other European policies, (3) changing the political framework for space in Europe, and (4) radical change—Community framework, substantial budget increase, thereby highlighting the impact of these options. Accordingly, the "no change" option would "imply abandoning the aim of bringing together the political commitment and actions of the EU, the EC, the ESA" (Ibid., p. 22) and the Member States. "There is a less pessimistic interpretation, which would assume that the current arrangements would stabilise and not be enhanced in the foreseeable future" (Ibid.), meaning that "the Commission and the ESA would continue to work closely together on programmes of mutual interest, particularly GALILEO" (Ibid.) and Copernicus, the

> Member States would conduct their own space programmes without any explicit coordination or exchange of information with European institutions and bodies, there would continue to be a fragmented institutional market facing European industry, and separate national positions would be taken vis-à-vis third countries, except possibly in the areas of the exploration initiative and the Group on Earth Observation. (Ibid.)

The second option of "Increased coordination and growing use of space applications to deliver other European policies" would contrary to the previous one "demonstrate a willingness to ensure that Europe's space activities must be cost-effective, well-coordinated and closely aligned to the needs of the operators and users" (Ibid.). Following this option the Commission and ESA would coordinate current and future programmes in this area. Further research needed to be done on the protection of space systems as critical infrastructures (Commission of the European Communities, 2007b, pp. 21–22). Following the third option "changing the political framework for space in Europe" increased coordination while retaining a non-EU intergovernmental framework for the majority of space activities in Europe, would bring intergovernmental activities into the EU framework. This way Member States could continue

> to participate optionally in programmes under intergovernmental funding arrangements. In parallel it would draw on the research and, as appropriate, operational budgets managed along Community lines. Suitable administrative arrangements would be necessary for non-EU Member States, which are currently members of ESA. (Ibid., p. 23)

A

> EU framework for European space activities could permit more effective coordination arrangements. It would be established in a way, which allowed expenditure under EU rules to be assured, without the need for negotiations between the Commission and the ESA on a case-by-case basis and without the need to establish new legal bodies programme by programme. (Ibid., p. 25)

The fourth option is the most radical one proposing to bring "space activities into the European Community framework within a clearly defined period as an integral part of a political decision to boost space investment significantly" (Ibid., p. 22).

> This would involve both the transfer of existing national civil space budgets into the Community budget and a commitment of new Community funds to space research or to operational space systems. (Ibid., p. 23)

and

> The radical option of itself would represent a substantial commitment of all Member States to a strengthened space investment, which would increase investor confidence in the industry. (Ibid., p. 26)

3.4.4 Commission Communication on an ESP: Framing Space in Strategic Terms

In April 2007, the Commission issued another Communication on the European Space Policy (cf. Commission of the European Communities, 2007a), which was welcomed and supported by a resolution adopted on 22 May 2007 by the Space Council.[41] The Commission's communication framed the space sector in strategic terms. Therein the development of a European Space Policy was considered "a strategic choice" for Europe, which will lead to a "strategic mission" in outer space. Space systems "are strategic assets demonstrating independence", which require "strategic investment decisions" (Commission of the European Communities, 2007a, p. 4). Tigner (2007) argues "One risks 'strategic' fatigue in reading the document, though it is only 17 pages long, including annexes." The Communication put forth twelve actions for the short term to support a European space policy, ranging from R&D proposals to funding for next-generation launcher technology over institutional reforms to proposals to support new integrated space applications (Ibid).

According to the Communication "[t]he EU will use its full potential to lead in identifying and bringing together user needs and to aggregate the political will in support of these and of wider policy objectives" (Commission of the European Communities, 2007a, p. 11), thereby ensuring the availability and continuity of operational services supporting its policies. ESA as well as its Member and Co-operating States are foreseen to "develop space technologies and systems, supporting innovation and global competitiveness and preparing for the future" (Ibid.), thereby focusing on exploration of space and on the basic tools: access to space, scientific knowledge and technologies. The EU is to "rely on the management and technical expertise of ESA, which will coordinate other relevant agencies and entities in Europe." (Ibid.) The FA is said to provide a "solid base for coordination arrangements between intergovernmental and Community actions" (Ibid., p. 12), having "provided significant

[41] This was a necessary step in order to make it a binding policy for the Member States. (cf. Mazurelle, Wouters, & Thiebaut, 2009, pp. 18–9).

advances in the working between the EC and ESA, and with the Member States, in policy development" (Ibid., p. 11). It will be assessed, improved if required (Ibid.), and

> complemented as needed based on an evaluation of experience to date. In addition, the Commission and ESA are conducting research on the main possible cost-efficient scenarios for optimising the organisation of space activities in Europe and adapting the EU-ESA relationship, in accordance with the request made by the second Space Council in June 2005. (Ibid., p. 14)

The European Economic and Social Committee (EESC) in response to the Communication advocated

> an open and transparent analysis and a dialogue on Europe's desirable performance in ten years' time (…) which objectives and corresponding institutional tools — regarding ESA, Commission and Member States — are needed to fulfil a jointly coordinated European mission. Amongst others, the dialogue should include the way ESA is financed, the dynamic contribution of medium-sized companies, and the maintenance of the highest level of competition. (European Economic and Social Committee, 2007a, p. 24)

While pointing to the success of the geo return, it fears "sudden changes of fixed procedures and relationships can be counter-productive" (Ibid.). The EESC's Consultative Commission for Industrial Change (CCMI) on European Space Policy provided a supplementary opinion (cf. European Economic and Social Committee, 2007b), proposing to renegotiate the FA "as to further strengthen links and relations between the EU and the ESA, while maintaining a clear line between respective competences and responsibilities." (p. 2). It further holds that "[a]s space policies become increasingly European in nature, cooperation between the EU and the ESA will need to be ever closer and more coordinated" (Ibid., p. 5).

The Communication's priorities were later to be reiterated by the EP in a resolution on how to bring space down to earth (cf. European Parliament, 2008).

3.4.5 *The Fourth Space Council: Concretising the Roles of EU and ESA*

On the basis of the Commission's Communication (Commission of the European Communities, 2007a) a jointly[42] prepared Resolution on the European Space Policy (cf. Council of the European Union, 2007), was adopted in May 2007 by the Council following its endorsement by the fourth "Space Council", inviting the European Commission and the ESA Executive "to establish a process of regular monitoring and priority setting" (Ibid., p. 11) through an implementation plan for the ESP. The Resolution on the European Space Policy was divided into two parts—the (1) vision for Europe and general strategy and (2) further steps—programmes (applications,

[42]The resolution was jointly drafted by the European Commission and ESA's Director General, Jean-Jacques Dordain and was adopted by the "Space Council".

security and defence, access to space, ISS and exploration, science and technology, governance, industrial policy, international relations) and Implementation.

While reaffirming the roles and responsibilities of the EU, ESA and Member States, as identified in the orientations of the second Space Council,[43] the resolution concretised the roles by proposing ESA's role should also include the support of the European Commission "as technical expert in the elaboration of European Community initiatives involving space-related activities and relevant work programmes, and in the selection and monitoring of relevant work contractors" and "the management by ESA of European Community space-related activities in accordance with the rules of the European Community" (Council of the European Union, 2007, pp. 8–9).

The resolution further

> calls on the European Commission to draw on the management and technical expertise of ESA for managing the European Community-funded R&D space infrastructure programmes with ESA coordinating the relevant agencies and entities in Europe. (Council of the European Union, 2007, pp. 8–9)

Apart from this, the fourth Space Council also identified four new priorities of the European Space Policy: space and climate change; contribution of space to the Lisbon strategy; space and security; and space exploration. It also highlighted Galileo and Copernicus as priorities. ESA DG Dordain referred to the resolution as an "important step forward for Europe", bringing "an EU dimension to space and a space dimension to the EU" (Tigner, 2007). "The U.S., Russia and China are moving fast in this sector and Europe needs to put its house in order with a coherent space policy," (Ibid.) he said during a press conference to unveil the policy.

3.4.6 *Taking Forward the ESP: Any Progress?*

The fifth Space Council, taking place on 26th September 2008, did not make any reference to governance questions in its Resolution entitled "Taking forward the European Space Policy" (cf. Council of the European Union, 2008). Instead it highlighted Galileo and Copernicus as the current priorities, and mentioned space and climate change, contribution of space to the Lisbon Treaty, space and security and space exploration[44] as the new priorities.

In September, the Commission had released a Space Policy Progress Report (cf. Commission of the European Communities, 2008), which touched upon the ESA-EU relationship in three sections. In the context of Galileo, the Commission held

> [p]ublic governance of the programmes shall be based on the principle of a strict division of responsibilities between the European Commission, the European GNSS Supervisory Authority (GSA) and ESA. The European Commission will be responsible for the overall management of the programmes as the European GNSS Programme Manager. ESA will

[43] i.e. ESA: development agency, EU: user needs and political will.

[44] For more information on Europe and space exploration cf. Landfester, Worms, Remuss and Schrogl (2010).

> act as procurement agent for the GALILEO Full Operational Capability (FOC), and as design agent for the programmes. The GSA, in accordance with guidelines issued by the Commission, will accomplish specific tasks related to the programmes. (Commission of the European Communities, 2008, p. 3)

On a more general level the Commission reiterated the division in optional and mandatory programmes (Ibid., 10–11). As key issues ahead, the Commission highlights "closer and more efficient cooperation [between EU and ESA], based on the FA, while exploring the need and mechanisms for future adaptations" (Ibid., p. 11). The progress report contained "Elements for a European Strategy for International Relations in Space" in its annex, which also refer to methodology and coordination mechanisms. Therein, the Commission summarised that "[t]he European Commission, ESA, EUMETSAT and Member States will exchange information and consult each other on ongoing and planned major international cooperation activities related to space." (Ibid., pp. 16–7). Additionally, the EC-ESA Joint Secretariat was foreseen to "provide for the necessary practical coordination, annual update, and access to the provided information, including through the setting up of an e-information-network for international relations" (Ibid., p. 16). Accordingly, coordinative actions will put Europe "in a better position to achieve jointly formulated objectives" (Ibid., p. 17).

3.4.7 The Economic Crisis: Putting Space in Perspective of Global Challenges

The ESA Ministerial Council (cf. ESA, 2008a) of 2008 is said to have "inject[ed] new impetus to ensure space's role as a key asset in facing global challenges" (ESA, 2008b). In details it recognised the "need to seek greater consistency of EU industrial and procurement rules with those of ESA allowing for the effective and transparent implementation of joint ESA and EU programmes" (ESA, 2008a, p. 5).

The economic crisis also affected the discussions in the space sector (Giannopapa, 2011 p. 71), resulting in the resolution "The Contribution of space to innovation and competitiveness in the context of the European Economic Recovery Plan, and further steps" (cf. Council of the European Union, 2009) issued by the sixth Space Council on 15th June 2009. The resolution does not tackle any governance or institutional related questions.

Also in 2009, the ESA Council issued a report on ESA's contribution to the structured dialogue on security and related activities (cf. ESA, 2009). Later that same year Commission President Barroso highlighted the need to protect space assets, when talking about "The Ambitions of Europe in Space" (Barroso, 2009).

3.5 Phase 5: The Lisbon Treaty: Introduction of Space as a Shared Competence (2009–2012)

The fifth phase of the development of space policy in Europe is characterised by the entering into force of the Lisbon Treaty in 2009, which granted shared competences in the field of space to the EU.

3.5.1 The Lisbon Treaty and Its Implications

The continuous push of the EU's Commission to introduce community competences for space in the founding treaties resulted in it being included in the Lisbon Treaty, which was signed by the heads of States and governments on 13th December 2007 in Lisbon. The Lisbon Treaty amends the Treaty on European Union (TEU), and the Treaty establishing the European Community (TEC), which is consequently, renamed Treaty on the Functioning of the European Union (TFEU). The Lisbon Treaty creates a legal basis for the action of the EU in certain areas not explicitly covered (Hesse & Hornung, 2012, p. 187), such as "space", using substantially the same wording as foreseen in the Treaty Establishing a Constitution for Europe of 2004 ("the Constitutional Treaty"). The Lisbon Treaty refers to "space" in two articles (cf. ESA, 2008c). First, in Article 4.3 of the TFEU:

> In the areas of research, technological development and space, the Union shall have competence to carry out activities, in particular to define and implement programmes; however, the exercise of that competence shall not result in Member States being prevented from exercising theirs.

Although this article safeguards the status quo of the balance of power between the EU, ESA and Member States, the EU is now not only specifically empowered to draft a European Space Policy but also allowed to "promote joint initiatives, support research and technological development and coordinate the efforts needed for the exploration and exploitation of space" through a European Space Programme (Mazurelle et al., 2009, pp. 21–29).

Secondly, in Article 189, included in the Title XIV of the TFEU headed "Research and technological development and space", it is said to:

> "1. To promote scientific and technical progress, industrial competitiveness and the implementation of its policies, the Union shall draw up a European space policy. To this end, it may promote joint initiatives, support research and technological development and coordinate the efforts needed for the exploration and exploitation of space.
>
> 2. To contribute to attaining the objectives referred to in paragraph 1, the European Parliament and the Council, acting in accordance with the ordinary legislative procedure, shall establish the necessary measures, which may take the form of a European space programme, excluding any harmonisation of the laws and regulations of the Member States.

3. The Union shall establish any appropriate relations with the European Space Agency.

4. This Article shall be without prejudice to the other provisions of this Title."

Together with Article 4(3), Article 189 of the Treaty enshrines the space policy as an EU shared policy. According to Article 189 space is a competence shared between the Member States and the EU, which is interpreted as a "parallel" competence. The scheme of "parallel" competences allows the Member States to implement their national space policies unaffected by eventual measure of the EU in the field of space (Hobe et al., 2006, p. 547). The EP in a later document clarifies: "the EU may act in parallel to Member States, without overcoming them (and ESA), but fostering stronger cooperation in areas where national industrial policy considerations might prevail today" (European Parliament, 2014, p. 44). The provisions of the Lisbon Treaty makes a clear attribution of competences to the EU in the space field, dissolving the uncertainties existing in the previous situation, by supporting EU competences in policy fields in which space based applications can serve as a tool. Article 189 (1) also provides a clear legal framework for the EU to coordinate efforts in exploration and exploitation of space. Article 189 (3) gives the EU a clear mandate in relation to ESA, emphasising the need for cooperation between both organisations in space, thereby neither predefining the nature of the relations, nor providing elements to qualify what "appropriate relations" with ESA imply. It thus remains consistent with the FA. Article 189 (1) also obliges the EU, to "draw up a European space policy". Although it does not explicitly mandate the Commission to do so, the Commission will do so given its right of initiative. ESA is tasked with elaborating and implementing a long-term European space policy (Article 2 of the ESA Convention). The use of the indefinite article "a" in both TFEU and ESA Convention could imply the coexistence of more than one policy; the word "European" is not defined in the treaties.

Previous documents have also touched upon this issue. Among these the FA foresaw the joint elaboration of the ESP, while the Communication on a European Space Policy foresaw the EU to elaborate a ESP and ESA to "develop space technologies and systems, supporting innovation and global competitiveness and preparing for the future" (Commission of the European Communities, 2007a, p. 11). Thus, the Lisbon Treaty further adds to the confusion about the roles of EU and ESA resulting in inefficiency, frictional losses and confusion (Rüttgers, 1989, p. 127). It does not provide guidance on the definition of appropriate instruments and mechanisms (e.g. funding sources) for the future actions of the EU concerning space. In fact, the provisions of the Lisbon Treaty need detailed implementation mechanisms, in particular for what concerns the relations with ESA.[45]

[45]This section is based on the following information documents ESA (2008c), European Commission (2009).

3.5.2 *Global Challenges and Benefits of Space Applications for European Citizens*

The discussion on the space also changed during the fifth phase. Space applications were put in the light of the benefits for citizens in terms of providing answers to global challenges. For example the seventh Space Council in its resolution "Global Challenges: Taking full Benefit of European Space Systems" (cf. Council of the European Union, 2010) of 25 November 2010 invited ESA, EU and Member States to further develop an overall space strategy, thereby referring to both the EU and ESA as key actors. It stressed that the relations "between the EU and ESA—based on clarity and complementarity of roles reflecting the respective strengths of both organisations—are the cornerstone of ensuring continued success of the European Space Policy and future endeavours in space" (Ibid., p. 12).

Similarly, in April 2011, the Commission issued Communication entitled "Towards a space strategy for the European Union that benefits its citizens" (cf. Commission of the European Communities, 2011). Therein space policy is understood as a response to the social, economic and strategic challenges for Europe. Space is conceived as an instrument to achieve the EU's internal and external policy goals, contributing to the satisfaction of the EU's social, economic and strategic needs (Sánchez Aranzamendi, 2011). Governance questions are tackled in a section entitled "Towards a well-structured governance", which reiterates the need to strengthen cooperation (Commission of the European Communities, 2011, p. 11). In a section entitled "Developing Relations between the EU and the ESA", the Communication also refers to the Lisbon Treaty and reiterates the previously suggested division of labour: "the EU should gather and identify the users' needs in order to ensure that space resources meet European citizens' needs in full" (Ibid., p. 12) and "the ESA (…) has strong technical and management infrastructure and could support the development of new space facilities as regards both intergovernmental and EU-funded programmes" (Ibid.). The Commission suggested that "ESA should continue to develop into an organisation with an intergovernmental and an EU dimension, in which military and civil programmes can coexist" (Ibid., p. 11). It should

> continue to develop management structures geared solely towards EU programmes. (…) A flexible membership structure should also be established in order to enable Switzerland and Norway to take part in some programmes, subject to an agreement with the EU. (Ibid., p. 11).

The EU or rather the Commission is positioned as a policy driver for the overall ESP, facilitating the coordination of national policies with the objective to make the utilisation of space capabilities in Europe more efficient. (Sánchez Aranzamendi, 2011, pp. 5–6)

3.5.3 Copernicus and Its Operations

In November 2011, the Commission released a Communication on Copernicus and its operations (cf. Commission of the European Communities, 2011b), which also contained provisions concerning governance, political supervision and management. Among other things, the Commission underlined the need for an appropriate governance scheme to be put in place by 2014. In details, the Communication foresaw that the Commission "should remain responsible (…) for the overall political coordination, including the negotiation of international agreements" (Commission of the European Communities, 2011b, p. 5). Therein the delegation of management of Copernicus to ESA is out ruled

> for several reasons: first, ESA is a research & development agency; second, ESA is a space agency whereas a large part of GMES goes beyond activities in space; and third, taking into account the fact that GMES should benefit citizens across the EU, the Commission favours a Community approach that involves all 27 Member States. (Ibid.)

Instead, as in the case with Galileo, "certain tasks relating to programme management such as evaluation, negotiation and follow up of contracts, could be delegated to the European Global Navigation Satellite System (GNSS) Agency (GSA)." (Ibid., pp. 5–6). This way even more power is given to the GSA instead of ESA, creating parallel competencies.

3.5.4 Establishing Appropriate Relations Between the EU and ESA

At the end of 2012 the Commission released a Communication entitled "Establishing appropriate relations between the EU and the European Space Agency" (cf. European Commission, 2012), which in line with the provisions contained in the Lisbon Treaty "outlines possible short and long term avenues for EU relations with ESA with a view to continuing an efficient cooperation" (Ibid., p. 3). It can be seen as a climax of the rapprochement of the EU and ESA, as it is the first document to explicitly name the issue at stake in its title.

The Commission referred to the "growing role of the EU in space" (Ibid., p. 2). Accordingly, "[a] closer relationship with the ESA would enable a further development of divisions of tasks" (Ibid.). It gave credit to ESA and the Member States for Europe's strong technological and industrial base and highlighted that it is heavily relying on "ESA's excellence" (Ibid.). The Commission underlined that "the emergence of space as an EU policy, the growing importance of EU space programmes and the EU's reliance on ESA's technical expertise have not yet translated into an evolution of the governance of space matters at European level." (Ibid.). Thereby five categories of structural obstacles in the current EU-ESA relations are highlighted: (1) mismatch of financial rules; (2) membership asymmetry; (3) asymmetry in security

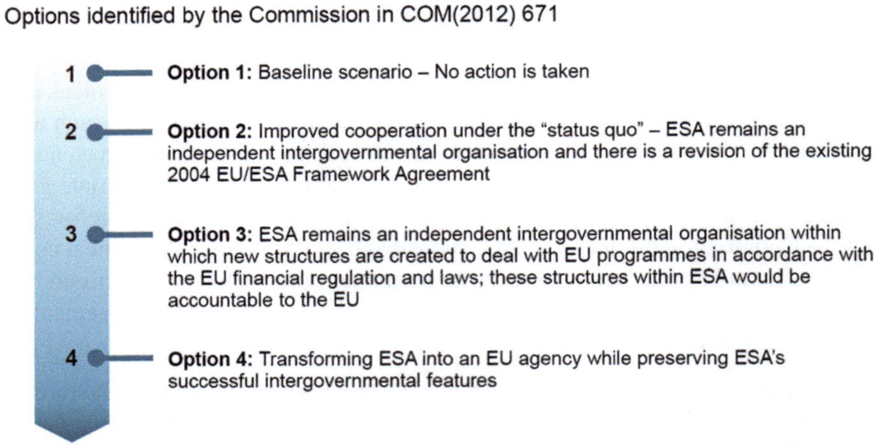

Fig. 3.2 Options identified by the Commission in COM (2012) 671 as summarised by Roland Berger in a Presentation to the Space Policy Expert Group in October 2013 (Roland Berger, 2013a, p. 3)

and defence matters; (4) absence of mechanisms for policy coordination; and (5) missing political accountability for ESA.

The Commission suggests for the EU to "provide political dimension (including at the international level) and legitimacy as well as links with other policy areas" as a "way ahead" (Ibid., p. 4). Accordingly, greater operational efficiency can only be achieved "through the rapprochement of ESA towards the European Union" (Ibid.). For this to be achieved, the Commission considers a target date between 2020 and 2025. It foresees to present possible options for such rapprochement before the end of 2013 (Ibid.). Among these options it mentions (s. also Fig. 3.2)

1. "bringing ESA as an intergovernmental organisation under the authority of the EU (following, to a certain extent, the model of the European Defence Agency)
2. transforming ESA into an EU agency (following the model of existing regulatory agencies).
3. Baseline scenario—no action is taken
4. Improved cooperation under the "status quo", incl. a revision of the FA." (Ibid., p. 4).

> These options would preserve the current essential features of ESA (i.e. optional programmes subscribed by Member States) while giving ESA key EU features – such as qualified majority decision-making or accountability vis-à-vis the European Parliament. (Commission of the European Communities, 2012, p. 4).

The Commission will accordingly "carry out a detailed cost-benefit and risk analysis of the different options, with a view to maximising synergies between the different actors including the GSA" (Ibid.) (Fig. 3.3).

Key principles of each option have been defined in order to evaluate options against identified limitations (1/2)

Overview of options

		OPTION 1	OPTION 2	OPTION 3	OPTION 4
Policy development	Definition of policy objectives	Commission and ESA in parallel	Commission and ESA jointly	Commission	Commission
	Definition of mission objectives	Commission and ESA in parallel	Commission and ESA jointly	Commission	Commission
	Funding of activities	Commission and ESA in parallel	Commission	Commission	Commission
Programme management	Arbitration in programme trade-offs (OTD vs. OTC vs. OTQ)	Commission	Commission	Commission, dir. EU Chamber	ESA as an EU agency
	Definition of technical design (Design Agent)	ESA	ESA	Commission, dir. EU Chamber	ESA as an EU agency
Project execution	Provision of technical expertise	ESA, partly duplicated at Commission	ESA, partly duplicated at Commission	ESA	ESA as an EU agency
	Responsibility for procurement (Contract Authority)	Commission, prepared by ESA	Commission, prepared by ESA	Commission, dir. EU Chamber	ESA as an EU agency

Fig. 3.3 Detailed overview of the options as identified by the Commission in COM (2012) 671 as summarised by Roland Berger in a Presentation to the Space Policy Expert Group in October 2013 (Roland Berger, 2013a, p. 4)

Building upon the previous Communication of April 2011 the Commission suggests, the EU could:

- "work through the Council of the EU, using as appropriate the open method of coordination to ensure coherence within the EU and consistency of EU Member States' positions in ESA with EU policies;
- Have systematic recourse to ESA for the design and development of EU space infrastructures, whereas exploitation activities will be carried out by other entities, such as the GSA;
- Ensure a homogeneous approach in line with the EU Financial Regulation for delegating responsibilities over EU space programmes to ESA as a way to prepare ESA to working within the EU environment, and for establishing partnerships with Member States when appropriate" (Ibid).

ESA in turn is advised to:

- "make the necessary structural adaptations (financial and internal decision-making) to ensure that activities delegated to ESA by the Commission are managed within an EU-like environment (e.g. through a dedicated directorate managing EU programmes within ESA);
- make the necessary changes allowing unrestricted access to the European Commission to ESA's relevant statutory bodies (e.g. ESA Council and its subordinate bodies) in order to give the Commission the possibility to provide input and ensure coordination with EU policies using existing mechanisms within ESA" (Ibid.).

3.5.5 Towards the European Space Agency that Best Serves Europe

On 20th and 21st November 2012 twenty ESA Member States governments, i.e. ministers in Naples—Italy, met to discuss the future of European space activities (Gruppe, 2012, p. 4). At the centre stage of discussions was among other things the future relationship between the EU and ESA (Wörner, 2012a), which following the latest Commission Communication were ever more undefined.

Preceding the actual conference, several meetings were held to reach agreement on disputed issues prior to the conference. Documents intended to show the way forward in the form of resolutions (ESA) and Communications (Commission) were published by both sides. According to Wörner (2012b), head of the German space agency DLR and chairman of the German delegation, "very dogmatic positions" were "being adopted and presented with great seriousness as the 'true solutions' for bringing both sides closer together" (p. 6). In contrast, "[a]n analytical consideration of the various stakeholders, of the legal bases of their documents and of their respective competencies does not (...) point in their stated directions" (Ibid.). Accordingly, both ESA and the EU Commissions have successfully demonstrated capabilities to justify their position as an actor in space policy. In this regard, ESA has a 50-years history as a European space agency under direct control of the Member States, with clear industrial policy requirements and an internal allocation of mandatory and optional programmes. At the same time the Commission has demonstrated, particularly in the field of research funding, that its mechanisms can be viable for Europe (Ibid.).

The result of the discussions in Naples was a political declaration entitled "Towards the European Space Agency that best serves Europe" (ESA, 2012), which aimed at setting the framework for any deliberations about the future of ESA within Europe (Kraft & Reutzel, 2012, p. 83). Accordingly, ESA is to evolve into a European Space Agency that will best serve Europe's position in a competitive environment, thereby taking full benefit and encompassing ESA's capabilities and achievements as well as the EU competence in space (ESA, 2012, pp. 2–3). The declarations does not go into details, as to how this should be done, but rather calls for proposals to be submitted at the next Council meeting at ministerial level, scheduled to take place in 2014. The general guideline is for European space activities to be conducted in a common understanding between Member States and the EU, with ESA being optimally engaged as a space agency to serve EU activities. Ministers promised to ensure coherence and coordination between the respective processes both on the ESA and EU side (Kraft & Reutzel, 2012, p. 83; ESA, 2012, p. 4). With the agreed declaration, the ministers confirm the intergovernmental basis of ESA also as the basis for its further evolution. The resolution is to emphasise the readiness of the ESA Member States to start a reflection on the further evolution of ESA, as to play their role in ensuring the coherence and coordination between both EU and ESA.

A special role in the process of achieving this document can be attributed to Germany. While it had taken a long time to arrive at a joint national position about what role the EU was to play alongside ESA, the German delegation "had made a special effort in drafting this paper" (Kraft, 2012, p. 17), with Wörner redrafting the paper in a nightshift to a degree that was finally acceptable to all Member States. In general, Germany does not support the existing efforts to integrate ESA into the EU, but considers an intergovernmental ESA to be necessary for a sustainable way of working. Accordingly, integration is not only "illogical but also unworkable". Wörner (2013) referred to the debate as a "cold potato" and "dead horse". He held that "the important advantages that ESA offers, particularly its direct control by representatives of the Member States, the optional programmes, and 'geographical return' as an instrument of national space policy" should not be lost. He classified the integration of ESA with the European Commission as suggest by the Commission (cf. Commission of the European Communities, 2011b) as "pointless". Thus, instead of continuing to discuss this option, he proposed to develop the details of the cooperation instead, thereby avoiding duplication. Special attention must be paid in this context to the GSA, which "should not become a new space agency, but must utilise ESA and its expertise".[46]

3.6 Phase 6: Post-ESA-EU Declaration of November 2012 (Post-2012)

3.6.1 Draft Council Conclusions "Establishing Appropriate Relations Between the EU and the European Space Agency"

Following the Commissions Communication of 14 November 2012 on "Establishing appropriate relations between the EU and the European Space Agency (ESA)" (cf. Commission of the European Communities, 2012), the Presidency presented Draft Conclusions on this Communication in January, which were then agreed upon as set out in the Annex of the Council's Conclusions entitled Draft Conclusions "Establishing appropriate relations between the EU and the European Space Agency" (cf. Council of the European Union, 2013). Therein, the Council acknowledged the important role "which ESA as an independent intergovernmental organisation plays in space matters and in relation to European space programmes, such as Galileo and COPERNICUS, together with other relevant actors" (Council of the European Union, 2013, p. 3). Additionally, it recognised a need

> to review and enhance the functioning of the relationship between the EU and ESA in view of the changes in the political context, the increasing role of the EU in the space domain, competitiveness challenges faced by the space sector and the growing importance of space activities for society (Ibid.),

thereby also acknowledging

[46]This section is based on Wörner (2013).

> that the "Framework Agreement" and its governance elements may, in its present form, no longer provide the most appropriate framework (Ibid.)

to ensure an efficient and effective European space policy fully utilising in a coherent manner the respective competencies of both the EU and ESA. It invited

> the European Commission to work together with the Director-General of ESA and in close cooperation with the respective EU and ESA Member States, including through existing forums[47] to ensure coherence, convergence and complementarity with a view to preparing **common proposals** on the further evolution of EU/ESA relations on the basis of common analysis, for decisions by the respective bodies by 2014. [emphasis added] (Ibid., p. 4)

3.6.2 Cost-Benefit Assessment of Options

On the basis of Draft Council Conclusions the Commission tasked Roland Berger, a strategy consulting firm, with the evaluation of EU-ESA relations in terms of a cost-benefit analysis of the four options as outlined by the Commission:

1. Baseline scenario—no action taken
2. Improved cooperation under the "status quo"; ESA remains an independent intergovernmental organisation; the EU/ES FAS of 2004 is to be reviewed
3. ESA remains an independent intergovernmental organisation within which new structure are created to deal with EU programmes in accordance with the EU financial regulation and laws. These structures within ESA would be accountable to the EU (i.e. EU pillar)
4. Transforming ESA into an EU agency while preserving ESA's successful intergovernmental features (Roland Berger, 2013b, p. 16).

Roland Berger's (RB) final study was expected to be published by September/October 2013. In the meantime, RB presented its project approach to the Space Policy Expert Group in Brussels on 18th July 2013.

Based on this, the Commission published a "Progress report on establishing appropriate relations between the European Union and the European Space Agency (ESA)" (cf. European Commission, 2014) in February 2014. The report "constitutes a step in the Commission's response to Council invitation (formulated in its conclusions of February 2013) to prepare proposals on the further evolution of EU/ESA relations" (European Commission, 2014, p. 9). Since the progress report is partly based on the Robert Berger external study, it reiterated the four options as highlighted by the Commission before. The Commission highlighted that these options have been addressed from, the perspective of their effectiveness and the costs associated with it, as well as a qualitative analysis of the ability of the option to resolve the identified structure limitation has been conducted.

The Commission concluded that "a conclusive choice among the options" (Ibid., p. 8) is not possible as the elements of analysis are "not (…) sufficient to justify"

[47] i.e. the High Level Space Policy Group and Space Policy Expert Group.

(Ibid.) such a choice. It quotes the Roland Berger Study, which concluded "the creation of a dedicated entity within ESA ...offers, in the medium term, the best compromise between expected effectiveness and ease of implementation" (Ibid.). As he Commission highlighted, it

> sees merits in this option, which needs to be further analysed and discussed with ESA in order to fully assess its legal, technical and human resource implications. The advantage of this option is that it would preserve the intergovernmental ESA structure, while creating an adequate framework to manage the financial resources brought by the EU. (Ibid.)

It further analysed that "While option 3 may be a better response to dealing with the structural issues in the EU/ESA relations, option 2 (the revision of the existing FA) should be considered alongside with option 3" (Ibid., p. 9).

The Commission emphasised it aims to "avoid a "big-bang" approach, but at the same time provide a solid and sustainable (legal) basis on which these relations can evolve in the long-term" (Ibid.).

As the next steps the Commission highlighted to "intensify discussions with Member States and with the Director-General of ESA" (Ibid.). It also indicated that "ESA is expected to take a decision about the evolution of the Agency during its Council meeting at ministerial level in December 2014" (Ibid.).

3.6.3 ESA's General Director Scenarios Assessment

As a consequence of the political declaration agreed upon in November 2012, ESA initiated a reflection process aimed at defining how the evolution of ESA can achieve the desired objectives (ESA, 2012, p. 3). To do so, ESA's Member States mandated ESA's DG Dordain to work with the EU Commission to provide an analysis of the situation of the European space sector. He was tasked to elaborate and assess in consultation with the ESA Council the different scenarios for ESA. In an interview on the document, he referred to the "continuous adaptation of ESA to a changing environment ever since its inception" (Pan European Network, 2013).

Based on these conclusions, ESA issued a so-called Action Plan for the Future Development of ESA (cf. ESA, 2013a) in March. Therein it is proposed to check four possible scenarios for their applicability for a future development of ESA:

- keeping the status quo
- development of an EU chamber within ESA, thereby keeping ESA as an intergovernmental organisation
- developing ESA into an agency such as the European Defence Agency (EDA), which is subordinated to the Council and is characterised by intergovernmental structures (cf. Bátora, 2009)
- developing ESA into an regulatory EU agency such as GSA, that is integrating ESA into the EU framework without keeping any intergovernmental characteristics (ESA, 2013a, p. 7).

The document proposed to analyse the four possible scenarios both from a cost-benefit point of view as well as by consulting ESA and EU Member States and European industry on the subject, as to determine ESA's future development from 2020 onwards (Ibid., p. 11).

As requested the DG coordinated his analysis to the extent possible with the European Commission, leading to a consistent set of scenarios to be assessed on both sides. As a result of this coordination process, the DG provided a document on the "Status of ESA/EU relations" (cf. ESA, 2013b), which he presented at the October Council. In contrast to RB's cost-benefit assessment he discussed three (instead of four) different possible scenarios:

(1) "Improved Status quo", in which the relationship between EU and ESA does not imply any change of the legal structure of each organisation.
(2) "EU Chamber", which implies adding a new structure to the current ESA legal structure, dedicated to EU funded programmes.
(3) "EU agency", which implies a redesign of the ESA legal structure.

The document presented in October contained a detailed description of the three scenarios. It was complemented by the ESA Council Working Paper of December 2013 entitled "Status of ESA/EU relations—Scenarios assessment" (ESA, 2013c), which presented a provisional roadmap for decisions to be taken during 2014. The document differentiates between short-, medium-, and long-term challenges for reaching the objectives set out in the political declaration of December 2012. It reiterated the scenarios as outlined before (ESA, 2013c, p. 5) and assessed them in the light of the following different criteria:

- related to efficient use of public funding and complementarity of activities
- regarding efficient articulation between space infrastructure and EU sectoral policies
- linked to impact on ESA programmes
- regarding impact on Member States
- regarding Impact on European Space industry activity
- regarding relations with Third Parties
- regarding efficiency and cost/savings if implementation and management
- regarding ease and rapidity of implementation (Ibid., pp. 28–9).

It was concluded that scenario 1 "has the potential to provide significant improvement, and provides the highest stability, but it requires a shared and balanced willingness between ESA and the EU to improve the current situation" (Ibid., p. 30). Scenario 2 was understood to provide "a similar level of improvement" (Ibid.), while bearing "some risk of degradation" (Ibid.) and representing "irreversible change" (Ibid.). Scenario 3 was outlined as "a high gain/high risk scenario" (Ibid.) providing "very little stability" (Ibid.), implying "significant change in terms of sources of funding and associated rules" (Ibid.) in particular industrial policy.

3.6.4 *"Space, Sovereignty and European Security—Building European Capabilities in an Advanced Institutional Framework"*

In January 2014, the EP's DG External Policies issued a study entitled "Space, sovereignty and European Security—Building European Capabilities in an advanced Institutional Framework" (cf. European Parliament, 2014). It aimed to offer a comprehensive analysis of the role of space-based capabilities in supporting the security and defence policies of the EU and its Member States. The first part covered the European space capabilities while the second part analysed the institutional framework, starting with the relevant elements of the Lisbon Treaty and continuing with an outline of the actors and roles in space related to security.[48] The document distinguished between upstream level and macro-roles, such as political guidance, priorities setting, budget approval and downstream level more general roles such as management, coordination and implementation of programmes. Accordingly, ESA is to be understood as "manager, developer and funder of the European launchers programs, but also the key European institutional customer" (European Parliament, 2014, p. 18).

Among its conclusions, the document referred to "the insufficient coordination among the key European space stakeholders" (Ibid., p. 74) which "continues to limit the efficiency of the whole European space sector" (Ibid.). It also highlighted "EU-ESA duplications and the absence of formal policy mechanisms to ensure that initiatives taken within ESA are consistent with EU policies" (Ibid.), which are said to represent "a weakness" (Ibid.). Among the difficulties to overcome this "weakness", it referred to "the mismatch of financial rules between ESA's geographic return principle and the EC competitive approach" (Ibid.) as well as the "asymmetries in terms of membership" (Ibid.). It also called for a "deeper integration of the activities undertaken by the EC, ESA, EEAS and EDA in the existing and diverse frameworks for dialogue and cooperation" (Ibid.), which should "accompany the ESA-EU rapprochement" (Ibid.). Accordingly a "definitive setting of the relationships between its main institutional actors—the EU and ESA (…) will take some time and will be the result of an extremely complex process" (Ibid.).

3.6.5 *Space Council Meetings in 2014*

The progress report provided the basis for the discussions about favourable scenarios within the Space Council in February 2014. The Council examined ways to improve cooperation between the organisations in order to prepare for forthcoming challenges in space policy projects. While several delegations pointed out that some of the scenarios presented in the Commission's progress report, or in combination with

---

[48] For further information on Europe as an actor in space and security cf. Rathgeber and Remuss (2009), Remuss (2010, 2011b).

other elements, could lead to more efficient cooperation, others felt the need for further cost-benefit analysis (Council of the European Union, 2014a). Romania was the only Member State pledging for scenario 1, i.e. keeping the status quo. Germany for example saw potential for improvement. Most Member States were in favour of either reviewing the FA or establishing a EU Chamber. Finally, the Presidency announced its intention to draft Council conclusions on the subject, based on the discussions and the Commission Progress Report, with a view to submitting them for adoption at the Competitiveness Council in May (Ibid.).

Consequently, the May Space Council adopted Conclusions entitled "Towards a shared EU-ESA vision for space fostering competitiveness" (cf. Council of the European Union, 2014b). These conclusions focused on the steps to be taken to provide proposals on improving the efficiency of relations between the EU and ESA, bearing in mind the broader space policy context and Europe's long-term strategic vision (Council of the European Union, 2014a). Therein, the Council invited the Commission and Member States "to work towards a reinforced partnership between the EU and ESA by exploiting their respective expertise and competencies in order to maximise efficiency and complementarity, while avoiding unnecessary duplication of activities" (Council of the European Union, 2014b, p. 5). Moreover, the Council acknowledged

> the important role of ESA as an independent intergovernmental organisation dedicated to space research and development of space systems, as well as its role in relation to Union space programmes together, where appropriate, with other relevant actors. (Ibid.)

In a similar vein it stressed

> that ESA constitutes an asset for Europe, in particular with regard to developing space technologies and systems, fostering Europe's access to space and pushing the frontiers of space-based scientific knowledge and research, supporting innovation and global competitiveness. (Ibid., p. 7)

It emphasised the essentiality of establishing

> appropriate, sustainable and transparent relations between the EU and ESA so as to maximise the impacts of European investment in space and to make the best use of competencies in Europe. (Ibid., p. 6)

The Council explicitly shared the Commission's assessment that transforming ESA into an EU agency would "require political consensus which may be difficult to reach in the foreseeable future" (Ibid., p. 7).

Since the May meeting of the EU Competitiveness Council the further analysis of scenario 4, i.e. the integration of ESA into the EU, has been postponed to an undefined time. Scenario 3 (EU chamber) and Scenario 2 (adaptation of the framework agreement) have been qualified. Further analysis and reasoning are asked for.

3.6.6 Position Paper of the Former Ministers Responsible for Space

Seven former Ministers, and, in one case, EU Commissioner and Parliamentarian who had the responsibility for space policy in their respective countries have published a position paper entitled "Bringing the Benefits of Space to the European Citizens—Innovation and Growth" (cf. Brinkhorst et al., 2014) in July 2014. The title already captures their main plea, which they referred to as a "renewed ambition". Accordingly, there is an "urgent need for political action in order to clarify the operational model of the EU-ESA institutional framework" (Ibid., p. 4). In their view the prime responsibilities of ESA and EU are both "different and complementary". As the prime responsibility of EU institutions, they see "to define EU sectoral policies and objectives and establish the requirements for the development of space technologies and infrastructure" as well as the "responsibility for the deployment of services required in order to attain its political goals" (Ibid.). "ESA, on the other hand" is said to have "full responsibility for the development of European space infrastructure (…) to meet the needs arising from European Union policies" as well as "the responsibility to provide scientific and technical leadership and advice in response to new opportunities for the deployment of EU policies made possible by scientific and technological progress". For the first time, they added another role for ESA "to continue to play a unique role for Europe, by attracting talent, resources and capabilities worldwide, able to strengthen EU international competitiveness" (Ibid., p. 5). Apart from defining mechanisms to ensure that ESA can contribute efficiently to EU policies the ministers also want to strengthen ESA by i.e. strengthening its constituencies: (1) ESA as the single agency for space research and development at European level, (2) ESA as an agency deriving full benefit from the flexibility of its structure and programme, (3) ESA as an agency with a strong industrial policy, (4) ESA as an agency fostering a model for global endeavours, (5) ESA as an independent agency.

3.6.7 The ESA Ministerial Conference of 2014

One of the topics under discussion at the ESA Ministerial Conference in December 2014 was the evolution of ESA, which was highlighted through a resolution (cf. ESA, 2014). While covering several aspects of ESA's evolution one chapter of the resolution deals with the "evolution of the relationship between ESA and the European Union." (ESA, 2014, p. 13) Therein the Member States' preferences "for a relationship between ESA and the European Union" are highlighted. Accordingly, the Member States want to keep "ESA as an independent, world-class intergovernmental space organisation according to the Convention" (Ibid.), and to make ESA "the long-term partner of choice for the EU for jointly defining and implementing the European Space Policy together with their respective Member States. This should be done in a dynamic and stepwise approach on the basis of the 2004 FA, with regular meetings of the Space Council" (Ibid., p. 14). The Member States further tasked the ESA DG "to continue the dialogue with the European Commission in order to maximise the coherence and complementarity between ESA activities and EU space-related activities, and development of agreed roadmaps for R&D" (Ibid.). They also tasked the ESA DG "to work with the European Commission for identifying and proposing, by 2016 to the ESA Council, improvements for developing a reliable and sustainable partnership between ESA and the EU" (Ibid.). Since DG Dordain retired in June 2015, it remains to be seen how the new German DG Wörner will influence the discussions.

References

Aliberti, M., & Krasner, S. D. (2016). Governance in space. In C. Al-Ekabi, B. Baranes, P. Hulsroj, & A. Lahcen (Eds.), *Yearbook on space policy 2014* (pp. 143–166). Vienna: Springer.

Aminzade, R. (1992). Historical sociology and time. *Sociological Methods & Research, 20,* 456–480.

Bailes, A. J. K. (2011). Death of an Institution. The end for Western European Union, a future for European defence? *Egmont Paper 46*. Retrieved from: http://aei.pitt.edu/32322/1/ep46.pdf. Accessed March 13, 2017.

Bainbridge, T. (2003). *The Penguin Companion to the European Union* (3rd ed.). London: Penguin Books.

Barroso, J. M. (2009). *Europe's ambitions in space*. Conference on European Space Policy. Brussels. Oct 15, 2009. Retrieved from http://europa.eu/rapid/press-release_SPEECH-09-476_en.htm. Accessed January 18, 2012.

Bátora, J. (2009). European Defence Agency: A flashpoint of institutional logics. *West European Politics, 32*(6), 1075–1098.

Battrick, B. (Ed.). (2003). *Agenda 2007—A document by the ESA Director General* (BR-213) Retrieved from: https://www.espi.or.at/images/documents/agenda%202007%20-%20a%20document%20by%20the%20esa%20director%20general%20.pdf. Accessed March 19, 2017.

Battrick, B. (Ed.). (2004). The European Space Sector in a global context—ESA's annual analysis (BR-222). Retrieved from: http://www.esa.int/esapub/br/br222/br222.pdf. Accessed March 19, 2017.

Bildt, C., Peyrelevade, J., & Späth, L. (2000). *Towards a space agency for the European Union*. [Wise-Men-Report]. Retrieved from: http://www.espi.or.at/images/documents/the%20wise%20men%20report.pdf. Accessed January 18, 2012.

Brachet, G. (2004). From initial ideas to a European Plan: GMES as an exemplar of European space strategy. *Space Policy, 20*, 7–15. 12.

Brinkhorst, L. J., Busquin, P., Gago, J. M., Profumo, F., Quilès, P., & Riesenhuber, H., et al. (2014). *Bringing the benefits of space to the European Citizens—Innovation and growth—A report to the Director General of European Space Agency by an independent group of former Ministers from ESA Member States*.

Busquin, P. (2001). Europe and space: Opening a new chapter. *Space Policy, 17,* 81–82.

Campbell, J. (2004). *Institutional change and globalization*. Princeton & Oxford: Princeton University Press.

Carpenter, A. (2016). Role of Galileo satellite technology in maritime security, safety and environmental protection. In T. Hörber & P. Stephenson (Eds.), *European Space Policy* (pp. 239–252). London & New York: Routledge.

Cheli, S., & Schrogl, K. (1999). Reshaping European space activities. *Space Policy, 15*(2), 61–66.

Commission of the European Communities. (1988). *The community and space—A coherent approach* (COM (88) 417). Retrieved from: http://aei.pitt.edu/3821/1/3821.pdf. Accessed March 13, 2017.

Commission of the European Communities. (1991). *The European Community. Crossroads in space. A report by an Advisory Panel* [Gibson Report]. Retrieved from http://www.google.de/url?sa=t&rct=j&q=&esrc=s&source=web&cd=2&cad=rja&uact=8&ved=0ahUKEwiJ7IuYjeLSAhUErRQKHYPKDZoQFggkMAE&url=http%3A%2F%2Fbookshop.europa.eu%2Fpl%2Fthe-european-community-pbCDNA14010%2Fdownloads%2FCD-NA-14-010-EN-C%2FCDNA14010ENC_001.pdf%3Bpgid%3Dy8dIS7GUWMdSR0EAlMEUUsWb0000GlIjFlwl%3Bsid%3Dps-HiJagmO-HwMQaDRkZL_SFoRyBWzRj2p8%3D%3FFileName%3DCDNA14010ENC_001.pdf%26SKU%3DCDNA14010ENC_PDF%26CatalogueNumber%3DCD-NA-14-010-EN-C&usg=AFQjCNG0PxmmE5-KunrpH2yXRwzuhTAgzA. Accessed March 13, 2017.

Commission of the European Communities. (1996). *The European Union and Space: Fostering applications, markets and industrial competitiveness* (COM (96) 617 final). Retrieved from http://aei.pitt.edu/3910/. Accessed March 19, 2017.

Commission of the European Communities. (1999). *Towards a coherent approach for space.* (SEC(1999) 189 final). Brussels: European Communities.

Commission of the European Communities. (2000). *Europe and space: Turning to a new chapter* (COM (2000)597 final). Retrieved from: http://eur-lex.europa.eu/legal-content/EN/TXT/PDF/?uri=CELEX:52000DC0597&from=EN. Accessed March 13, 2017.

Commission of the European Communities. (2001). *Towards a European Space Policy* (COM(2001)718 final). Retrieved from: http://cordis.europa.eu/publication/rcn/1740_en.html. Accessed March 13, 2017.

Commission of the European Communities. (2003a). *Space: A New European Frontier for an Expanding Union—An action plan for implementing the European Space Policy (COM (2003) 673) [White Paper]*. Brussels: European Communities.

Commission of the European Communities. (2003b). Green Paper on European Space Policy—Discussion point: The institutional framework of a New European Space Policy. Retrieved from: http://studylib.net/doc/10959644/green-paper-on-european-space-policy. Accessed March 13, 2017.

Commission of the European Communities. (2005). *European Space Policy—Preliminary elements (COM (2005) 208 final)*. Brussels: European Communities.

Commission of the European Communities. (2007a). *European Space Policy* (COM (2007) 212 final). Retrieved from http://eur-lex.europa.eu/legal-content/EN/TXT/?uri=CELEX:52007DC0212. Accessed March 13, 2017.

Commission of the European Communities. (2007b). *Impact assessment of the European Space Policy (SEC (2007) 505 draft)*. Brussels: European Communities.

Commission of the European Communities. (2012). *Establishing appropriate relations between the EU and the European Space Agency (COM(2012) 671 final)*. Retrieved from: http://eur-lex.europa.eu/legal-content/EN/TXT/PDF/?uri=CELEX:52012DC0671&from=EN. Accessed March 13, 2017.

Commission of the European Communities. (2008). *European Space Policy progress report (COM (2008) 561 final)*. Brussels: European Communities.

Commission of the European Communities. (2011a). *Towards a space strategy for the European Union that benefits its citizens (COM (2011) 152 final)*. Brussels: European Communities.

Commission of the European Communities. (2011b). *On the European Earth monitoring programme (GMES) and its operations (from 2014 onwards) (COM (2011) 831 final)*. Brussels: European Communities.

Conner, A. (2010). Introduction: International institutions and effectiveness—Achieving cooperation. In J. L. Goldstein & R. H. Steinberg (Eds.), *International Institutions—Volume II consequences: When, where and why international institutions are effective* (pp. iiix–x). London: Sage.

Council of the European Union. (1999). *Council Resolution on developing a coherent European space strategy* (1999/C 375/01). Retrieved from: http://eur-lex.europa.eu/legal-content/EN/TXT/?uri=CELEX%3A31999Y1224(01). Accessed March 13, 2017.

Council of the European Union. (2000). *Resolution on a European Space Strategy (2000/C 371/02).* Brussels: European Communities.

Council of the European Union. (2003). *Council Decision on the signing of the Framework Agreement between the European Community and the European Space Agency* (12858/03). Brussels: European Communities. Retrieved from: http://ec.europa.eu/DocsRoom/documents/2894/attachments/1/translations/en/renditions/native. Accessed March 25, 2017.

Council of the European Union. (2004). *Report on the First Space Council (15000/04 LIMiTE RECH 234).* Brussels: European Communities.

Council of the European Union. (2005a). *Report on the Second Space Council (9440/05 RECH).* Brussels: European Communities.

Council of the European Union. (2005b). *Report on the Third Space Council* (14499/1/05 REV 1). Brussels: European Communities. Retrieved from: http://register.consilium.europa.eu/doc/srv?l=EN&f=ST%2014499%202005%20REV%201. Accessed March 25, 2017.

Council of the European Union. (2006). *Council Joint Action amending Joint Action 2001/554/CFSP on the Establishment of a European Union Institute for Security Studies (2006/1002/CFSP).* Brussels: European Communities.

Council of the European Union. (2007). *Resolution on* the *European Space Policy* (10037/07). Brussels: European Communities. Retrieved from: http://register.consilium.europa.eu/doc/srv?l=EN&f=ST%2010037%202007%20INIT. Accessed March 13, 2017.

Council of the European Union. (2008). *Resolution on taking forward the European Space Policy (13569/08).* Brussels: European Communities.

Council of the European Union. (2009). *Resolution on the contribution of space to innovation and competitiveness in the context of the European Economic Recovery Plan, and further steps (10500/09).* Brussels: European Communities.

Council of the European Union. (2010). *Global challenges: Taking full benefit of European space systems (Resolution 16864/10).* Brussels: European Communities.

Council of the European Union. (2013). *Draft Conclusions "Establishing appropriate relations between the EU and the European Space Agency" (6183/13).* Brussels: European Communities.

Council of the European Union. (2014a). *Press Release—3295th Council meeting competitiveness (internal market, industry, research and space)—Main conclusions.* Retrieved from: www.consilium.europa.eu/uedocs/cms_data/docs/pressdata/en/intm/141115.pdf. Accessed March 13, 2017.

Council of the European Union. (2014b). *Towards a shared EU-ESA vision for space fostering competitiveness—Draft Council conclusions* (9851/14 of 20 May 2014). Retrieved from: http://register.consilium.europa.eu/doc/srv?l=EN&f=ST%209851%202014%20INIT. Accessed March 13, 2017.

Davara, F. (2000). The future of the WEU Torrejon Centre. *Air & Space Europe, 2*(4), 24–25.

De Maria, M. (1993). *The history of ELDO Part 1: 1961–1964.* ESA History Study Report HSR-10. Nordwijk: ESA.

Dudley, G., & Richardson, J. (1999). Competing advocacy coalitions and the process of "frame reflection": A longitudinal analysis of EU steel policy. *Journal of European Public Policy, 6*(2), 225–248.

Dupas, A., Janischewski, S., von Kries, W., & Schrogl, K. (2001). A Franco-German view of Europe's ambition in space for the 21st century. *Space Policy, 17,* 103–110.

ESA. (1975). *Convention for the establishment of a European Space Agency.* Retrieved from: http://www.esa.int/About_Us/Law_at_ESA/ESA_Convention. Accessed March 19, 2017.

ESA. (1985). *Council Resolution on the long-term European space plan* (ESA/C/LXVII/res1 (final)). Retrieved from: http://download.esa.int/docs/LEX-L/Ministerial-Council-Resolutions/Meeting-C-M067(Rome,30–31Jan1985)/ESAC-MLXVIIRes.1(Final).pdf. Accessed March 19, 2017.

ESA. (1998a). *ESA Bulletin 95*. Retrieved from http://www.esa.int/esapub/bulletin/bullet95/COUNCIL.pdf. Accessed March 19, 2017.

ESA. (1998b). *Resolution on the reinforcement of the synergy between the European Space Agency and the European Community* (ESA/C/CXXXVI/Res.1 (final)). Paris: ESA.

ESA. (2000). *Resolution on a European strategy for space* (ESA/C-M/CXLVIII/Res.1 (final) bzw. ESA/C(2000)67 rev.1). Retrieved from: http://download.esa.int/docs/LEX-L/Ministerial-Council-Resolutions/Meeting-C-M148(Brussels,16Nov2000)/ESAC-MCXLVIIIRes.1(Final).pdf. Accessed March 19, 2017.

ESA. (2007a). European Space Agency. "Agenda 2011." *ESA Bulletin, 128*, pp. 8. Retrieved from http://www.esa.int/esapub/bulletin/bulletin128/bul128a_agenda2011.pdf. Accessed March 19, 2017.

ESA. (2008a). *Entschließung über die Rolle der Raumfahrt bei der Verwirklichung der globalen Ziele Europas* (ESA/C-M/CCVI/Res.1 (final)). Retrieved from http://www.espi.or.at/images/stories/ESA_2008_Council_Meeting.pdf. Accessed March 19, 2017.

ESA. (2008b). *European Ministers inject new impetus to ensure space's role as a key asset in facing global challenges* (Press Release 47–2008). Retrieved from: http://www.esa.int/For_Media/Press_Releases/European_Ministers_inject_new_impetus_to_ensure_space_s_role_as_a_key_asset_in_facing_global_challenges. Accessed March 19, 2017.

ESA. (2008c). *Space in the "Lisbon Treaty"—Information document* (ESA/C(2008)19). Retrieved from: http://www.espi.or.at/images/stories/ESA_Space_in_the_Lisbon_Treaty.pdf. Accessed March 19, 2017.

ESA. (2009). *ESA's contribution to the structured dialogue on security and related activities (ESA/C(2009) 12)*. Paris: ESA.

ESA. (2012). *Political declaration towards the European Space Agency that best serves Europe* (ESA/C-M/CCXXXIV/Res.4 (final)). Naples: ESA.

ESA. (2013a). *Aktionsplan für die Weiterentwicklung der ESA (ESA/C(2013) 9)*. Paris: ESA.

ESA. (2013b). *Status of ESA/EU relations (ESA/C(2013) 94)*. Paris: ESA.

ESA. (2013c). *Status of ESA/EU relations—Scenarios assessment (ESA/C(2013) 139)*. Paris: ESA.

ESA. (2014). *Resolution on ESA evolution* (ESA/C-M/CCXLVII/Res. 3 (Final)). Retrieved from http://esamultimedia.esa.int/docs/corporate/Final_resolutions_1_2_3_from_CM_2014_Releasable_to_the_public.pdf. Accessed March 19, 2017.

European Commission. (2009). *Information Note "Article 189 of the Treaty on the Functioning of the EU." (HSPG/22-2009)*. Retrieved from http://www.espi.or.at/images/stories/HSPG_Information_Note.pdf. Accessed March 17, 2017.

European Commission. (2012). *Establishing appropriate relations between the EU and the European Space Agency* (COM (2012) 671 final). Brussels: European Communities.

European Commission. (2014). *Progress report on establishing appropriate relations between the European Union and the European Space Agency (ESA)* (COM (2014) 56 final). Brussels: European Communities.

European Economic and Social Committee. (2007a). *Opinion of the European Economic and Social Committee on the COM (2007) 212 final*. OJ C 126 of 25 June 2008. Brussels: European Communities.

European Economic and Social Committee. (2007b). *Supplementary Opinion of the Consultative Commission for Industrial Change (CCMI) on European Space Policy*. CCMI/048 of 3 December 2007. Brussels: European Communities.

European Parliament. (1979). *Resolution on Community participation in space research* [Ripamonti Report]. OJ C 127. Retrieved from http://eur-lex.europa.eu/LexUriServ/LexUriServ.do?uri=OJ:C:1979:127:FULL:EN:PDF. Accessed March 19, 2017.

European Parliament. (1981). *Resolution on European Space Policy* [Turcat Report]. OCJ C 260.

European Parliament. (1987). *Resolution on European Space Policy*. OJ C190. http://eur-lex.europa.eu/legal-content/EN/TXT/?uri=CELEX:51987IP0066. Accessed March 10, 2015.

European Parliament. (1991a). Rovsing report. Draft of March 1991, PE-146.210/Corr. Draft of 6 March 1991 and Final Report of 15 October 1991 bearing the same number.

European Parliament. (1991b). *Resolution on European Space Policy*. OJ C 305. Retrieved from http://publications.europa.eu/resource/cellar/9b45eb8d-1a58-4568-baf8-557aacabdeda.0004.01/DOC_1. Accessed March 19, 2017.

European Parliament. (1994). *Community and space*. Retrieved from http://eur-lex.europa.eu/LexUriServ/LexUriServ.do?uri=OJ:C:1994:205:FULL:EN:PDF. Accessed March 19, 2017.

European Parliament. (1998). *Resolution of 13 January 1998 on the Commission communication to the Council and Parliament entitled 'The European Union and space: fostering applications, markets and industrial competitiveness'* (COM (06) 617 – C4-0042/97). OJ C 34. Retrieved from http://eur-lex.europa.eu/LexUriServ/LexUriServ.do?uri=CELEX:51997IP0384:EN:HTML. Accessed March 19, 2017.

European Parliament. (2003). *Report on European Space Policy—Green Paper* (2003/2092(INI). Retrieved from http://www.europarl.europa.eu/sides/getDoc.do?type=REPORT&mode=XML&reference=A5-2003-294&language=EN. Accessed March 19, 2017.

European Parliament. (2008). *Resolution on the European space policy: How to bring space down to earth.* Strasbourg: European Parliament.

European Parliament. (2014). *Space, sovereignty and European Security—Building European Capabilities in an advanced Institutional Framework* (EXPO/B/SEDE/2012/21). Retrieved from: http://www.europarl.europa.eu/RegData/etudes/etudes/join/2014/433750/EXPO-SEDE_ET(2014)433750_EN.pdf. Accessed March 19, 2017.

Froehlich, A. (2008). Die European Space Policy und ihre Auswirkungen auf die Verlängerung des ESA/EG Rahmenabkommens. *Zeitschrift für Luft- und Weltraumrecht (ZLW), 57*(1), 67–77.

Giannopapa, C. (2011). The less known, but crucial elements of the European Space Flagship Programmes: Public perception and international aspects of Galileo/EGNOS and GMES. ESPI Report 34. Vienna: ESPI.

Gibbons, M. A. (1986). Managing cooperation and conflict within the European Space Agency. *Annals of International Studies, 15,* 27–44.

Gruppe, G. (2012). Editorial. *DLR Newsletter Countdown*. Special Edition of December 2012. Bonn: DLR.

Hall, P. A., & Taylor, R. C. R. (1996). *Political science and the three new institutionalisms* (pp. 936–957). XLIV: Political Studies.

Hesse, M., & Hornung, M. (2012). Neue Impulse der Europäischen Kommission für eine europäische Weltraumstrategie. *Integration, 35*(3), 183–199.

Hobe, S. (2004). Prospects for a European Space Administration. *Space Policy, 20,* 25–29.

Hobe, S., Heinrich, O., Kerner, I., & Froehlich, A. (2009). *Entwicklung der Europäischen Weltraumagentur als „implementing agency" der Europäischen Union: Rechtsrahmen und Anpassungserfordernisse*. Kölner Schriften zum Internationalen und Europäischen Recht. Band 17. Berlin: LIT Verlag.

Hobe, S., Kunzmann, K., & Reuter, T. (2006). *Forschungsbericht ESA – EU: Rechtliche Rahmenbedingungen einer zukünftigen kohärenten Struktur der europäischen Raumfahrt*. Kölner Schriften zum Internationalen und Europäischen Recht Bd. 13. Berlin: LIT Verlag.

Hörber, T. (2009). ESA + EU: Ideology or pragmatic task sharing? *Space Policy, 25,* 206–208.

Hörber, T. (2016a). The European Space Agency and the European Union. In T. Hörber & P. Stephenson (Eds.), *European Space Policy* (pp. 53–65). London & New York: Routledge.

Hörber, T. (2016b). Conclusion: "To boldly go where no one has gone before". In T. Hörber & P. Stephenson (Eds.), *European Space Policy* (pp. 253–263). London & New York: Routledge.

Hörber, T. (2016). Foreword. In T. Hörber & P. Stephenson (Eds.), *European Space Policy* (pp. xx–xxiii). London & New York: Routledge.

Hörber, T. (2016d). Introduction—Towards a European Space Policy. In T. Hörber & P. Stephenson (Eds.), *European Space Policy* (pp. 1–12). London & New York: Routledge.

Hörber, T. (2016e). Chaos or consolidation? Post-war space policy in Europe. In T. Hörber & P. Stephenson (Eds.), *European Space Policy* (pp. 15–29). London & New York: Routledge.

Hutter, T. (1991). *Charakteristika und Bestimmungsfaktoren der europäischen Raumfahrtpolitik*. Dissertation zur Erlangung des akademischen Grades Doktor der Sozialwissenschaft an der Fakultät für Sozial- und Verhaltenswissenschaften der Eberhard-Karls-Universität Tübingen.

Kraft, G. (2012). *For one German opinion*. DLR Newsletter Countdown. Special Edition of December 2012. Bonn: DLR.

Kraft, G., & Reutzel, I. (2012). Political resolutions. *DLR Newsletter Countdown*. Special Edition of December 2012. Bonn: DLR.

Köpping Anthanasopoulos, H. (2016). Europe's wilderness—The Council's frames on space policy. In T. Hörber & P. Stephenson (Eds.), *European Space Policy* (pp. 82–97). London & New York: Routledge.

Krige, J. (1992). *The prehistory of ESRO 1959/60 from the first initiatives to the formation of the COPERS*. ESA History Study Report HSR-1. Nordwijk: ESA.

Krige, J. (1993). *Europe into space: The Auger Years (1959–1967)*. ESA History Study Report HSR-3. Nordwijk: ESA.

Krige, J. (1993b). *The early activities of the COPERS and the drafting of the ESRO Convention (1961/62)*. ESA History Study Report HSR-4. Nordwijk: ESA.

Krige, J. (1993c). *The launch of ELDO*. ESA HSR-7. Nordwijk: ESA.

Krige, J. (1994a). The European space system. In J. Krige & A. Russo (Eds.), *Reflections on Europe in space* (pp. 1–13). ESA HSR-11. Nordwijk: ESA.

Krige, J., & Russo, A. (2000). *A history of the European Space Agency: 1958–1987*. Volume I. SP - 1235 of April 2000. Nordwijk: ESA.

Krige, J., Russo, A., & Sebesta, L. (2000). *A history of the European Space Agency: 1958–1987*. Volume II. SP - 1235 of April 2000. Nordwijk: ESA.

Kunzmann, K., & Reuter, T. (2004). Crafting a legal framework for a coherent future structure for European Space Activities. *Space Policy, 20*, 59–61.

Landfester, U., Worms, J.-C., Remuss, N.-L., & Schrogl, K.-U. (Eds.). (2010). *Humans in outer space—Interdisciplinary perspectives*. Vienna: Springer.

Lüst, R. (1987). *Die Zukunft der europäischen Raumfahrt*. Melle: Knoth.

Lüst, R. (1989). Die Europäische Weltraumorganisation ESA – Geschichte, Aufbau und Langzeitplan. In J. Rüttgen (Ed.), *Europas Wege in den Weltraum: Programme – Protest – Prognosen*. Frankfurt am Main (pp. 147–63). Umschau Verlag.

Massey, H. (1986). *History of British space science*. Cambridge: Cambridge University Press.

Madders, K. (1997). *A new force at a new frontier—Europe's development in the space field in the light of its main actors, policies, law and activities from its beginning up to the present*. Cambridge: Cambridge University Press.

Madders, K., & Thiebaut, W. M. (1992). Two Europe in one space: The evolution of relations between the European Space Agency and the European Community in Space Affairs. *Journal of Space Law, 20*(2), 117–132.

Madders, K., & Wouters, J. (2004). Taking stock of Europe's developing space policy: From the European space policy workshops to the European space policy forum. *Space Policy, 20*, 31–36.

Marchisio, S. (2006). Proposals for an institutional realignment of the European Space Sector. In S. Hobe, B. Schmidt-Tedd, & K.-U. Schrogl (Eds.), *Project 2001 Plus—Global and European challenges for air and space law* (pp. 193–204). Köln: Heymanns.

Marta, L., & Stephenson, P. (2016). Role of the European Commission in framing European Space Policy. In T. Hörber & P. Stephenson (Eds.), *European Space Policy* (pp. 98–113). London & New York: Routledge.

Mazurelle, F., Wouters, J., & Thiebaut, W. (2009). *The evolution of European Space Governance: Policy, legal and institutional implications*. Working Paper No. 25 (2009). Leuven: Leuven Centre for Global Governance Studies. Retrieved from https://www.law.kuleuven.be/iir/nl/onderzoek/wp/wp135e.pdf. Accessed March 19, 2017.

Micklitz, H. W., & Reich, N. (1989). *Legal aspects of European space activities—ESA convention, EEC internal market and common commercial policy*. Baden-Baden: Nomos.

Pan European Network. (2013). *ESA's 2012 Council at Ministerial Level*. Retrieved from: http://www.paneuropeannetworks.com/interview/esas-2012-council-at-ministerial-level/. Accessed March 29, 2017.

Peter, N. (2007). The EU's emergent space diplomacy. *Space Policy, 23,* 97–107.

Rathgeber, W., & Remuss, N. (2009). *Space security—A formative role and principled identity for Europe*. ESPI Report 16. Wien: ESPI.

Remuss, N. (2010a). Space and security—Challenges for Europe. *Sicherheit und Frieden, 3,* 151–160.

Remuss, N. (2010b). Space applications as a supporting tool for countering piracy—Outline for a European approach. *ESPI Report 29*. Vienna: ESPI.

Remuss, N. (2010c). Space and maritime security—Strategies for countering piracy. *Space Policy, 26*(2), 124–125.

Remuss, N. (2011a). Meeting Europe's external and internal security through space applications—Security as an element for European identity. In B. Baranes & K.-U. Schrogl (Eds.), *European identity through space* (pp. 132–44). Vienna: Springer.

Remuss, N. (2011b). Space and security. In C. Brünner & A. Soucek (Eds.), *Outer space in society, politics and law* (pp. 519–554). Vienna: Springer.

Remuss, N. (2011c). The use of space resources in the fight against piracy. *Journal of JAPCC, 13,* 37–40.

Remuss, N. (2011d). The use of space resources in the fight against piracy. *European Journal of Navigation, 9*(1), 19–26.

Remuss, N. (2014). The use of space resources in the fight against piracy. *NMIOTC Journal, 8,* 28–40.

Reuter, T. (2007). *Die ESA als Raumfahrtagentur der Europäischen Union - Rechtliche Rahmenbedingungen für eine institutionelle Neuausrichtung der europäischen Raumfahrt*. Köln: Carl Heymanns Verlag.

Roland Berger. (2013a). *Evolution of EU-ESA relations: Cost/benefit assessment of options*. Presentation of the project outcome to the Space Policy Expert Group. Brussels, 30.10.2013.

Roland Berger. (2013b). *Evolution of EU-ESA relations. Cost/benefit assessment of options*. Work Package 1. Brussels.

Rüttgers, J. (1989). *Europas Wege in den Weltraum: Programme – Proteste – Prognosen*. Frankfurt am Main: Umschau Verlag.

Sánchez Aranzamendi, M. (2011). European Commission Communication "Towards a space strategy for the European Union that benefits its citizens"—Towards a Lisbon Generation Competence? *ESPI Perspectives, 46*. Retrieved from: http://www.espi.or.at/images/stories/dokumente/Perspectives/ESPI_Perspectives_46.pdf. Accessed March 13, 2017.

Schrogl, K. (1993). Die Europäische Gemeinschaft als Magnet – Ihre Anziehungskraft auf die europäischen Fachorganisationen ESA und CEPT. *Zeitschrift für Internationale Politik, 18,* 525–532.

Sheehan, M. (2007). *The international politics of space*. London: Routledge.

Silvestri, S. (2003). *Space security policy in Europe*. Paris: European Institute for Security Studies.

Smith, L. J., & Hörl, K. U. (2007). Institutional challenges for space activities in Europe. *Acta Astronautica, 60,* 210–220.

Suzuki, K. (2003). *Policy logics and institutions of European space collaboration*. Burlington: Ashgate.

The European Convention Secretariat. (2003). Draft Treaty establishing a Constitution for Europe. Brussels: European Communities. Retrieved: http://european-convention.eu.int/docs/treaty/cv00850.en03.pdf. Accessed March 29, 2017.

Thelen, K. (1999). Historical institutionalism in comparative politics. *Annual Review of Political Sciences, 2,* 369–404.

Tigner, B. (2007). *EU, ESA: In search of a common space policy*. ISN Security Watch, 3rd May 2007. Retrieved from: http://www.isn.ethz.ch/Digital-Library/Articles/Detail/?ots783=4888caa0-b3db-1461-98b9-e20e7b9c13d4&lng=en&id=53222. Accessed October 14, 2008.

Times Higher Education. (2004). New Constitution gives green light to EU space activities. Times Higher Education, 14 July 2004. Retrieved from: https://www.timeshighereducation.com/news/new-constitution-gives-green-light-to-eu-space-activities/190137.article. August 12, 2016.

Verheugen, G. (2005). Europe's space plans and opportunities for cooperation. *Space Policy, 21,* 93–95.

von Kries, W. (2003). Which future for Europe's space agencies? *Space Policy, 19,* 157–161.

Wörner, J. (2012a). A heated debate concerning 'a cold potato.' *DLR Blogs – Jan Woerner's Blog*. Retrieved from http://www.dlr.de/blogs/en/archive/jan-woerner/a-heated-debate-concerning-a-cold-potato.aspx. Accessed March 18, 2017.

Wörner, J. (2012b). There was much at stake. *DLR Newsletter Countdown*. Special Edition of December 2012. Bonn: DLR.

Wörner, J. (2013). Von der kalten Kartoffel zum toten Pferd …. DLR Blogs – Jan Woerner's Blog. Retrieved from: http://www.dlr.de/blogs/de/desktopdefault.aspx/tabid-5896/9578_read-632/. Accessed March 18, 2017.

Chapter 4
Theoretical Discussion

This dissertation has first reviewed the state of the art of literature dealing with institutional change and the European integration process. The second chapter thus gave an overview of the applicable tenants of European Integration Theories (i.e. Neofunctionalism, Intergovernmentalism and Liberal Intergovernmentalism, Supranational Governance), Agency Theories, Theories of International Relations and New Institutionalism. In line with the second movement in institutional analysis of the 1990s (Campbell, 2004, p. 4), different theories were understood to complement each other. On this basis, five stages of institutional development and five facets of institutional change were identified to structure the analysis of institutional change and the observations of the case under discussion. In line with Keohane and Nye (2001, p. 50), these were perceived as most important in describing institutional change.

In a second step the historic development of space policy in Europe has been outlined from the beginning of European space collaboration in 1959 until 2014. A chronological approach has been adopted. Bearing in mind the importance of the time frame for the analysis of institutional change, six phases of institutional development of space policy in Europe have been identified, which in line with Aminzade (1992) and Campbell (2004, pp. 44–6) were distinguished by critical events.[1] This last chapter seeks to combine the theoretical and empirical part of this dissertation. To do so, the six phases of the development of space policy in Europe will be analysed in terms of the stages of institutional development and the five facets of institutional change. Particular attention is paid to the impact of the European integration process.

[1] Accordingly, critical events may mark the appropriate time frames for analysis (e.g. electoral, budget, legislative or other cycles).

N.-L. Remuss, *Theorising Institutional Change: The Impact of the European Integration Process on the Development of Space Activities in Europe*, Springer Theses, https://doi.org/10.1007/978-3-319-95978-8_4

4.1 Stages of Institutional Development and the Development of Space Policy in Europe

This dissertation perceives institutions in line with Sandholtz & Stone Sweet (1998) as being in constant evolution (Sandholtz & Stone Sweet, 1998, p. 16). Building on Levy, Young, and Zürn (1995) and Florensa (2004), the process of institutional development is understood to consist of the following five stages:

- agenda formation (Levy et al., 1995),
- institutional choice (i.e. selection and design) (Ibid.),
- operationalisation (Ibid.) or evolution (Florensa, 2004)
- redefinition (Ibid.) and
- crisis (Ibid.).

Considering the case study at hand, the question arises, which stages of institutional development ESA went through in the identified six phases of institutional development of space activities in Europe. The case of ESA might also give insights to the conditions for the foundation of an institution, which have been identified based on Rittberger and Zangl (2006, p. 24), DiMaggio (1988) and Olsen (1997, pp. 209–10, 217) as the "problem condition", the "cognitive condition", the "hegemonic condition" (Rittberger & Zangl, 2006, p. 16), "windows of opportunity", and breaking points.

Moreover, the case study might also give insights into the reasons for the selection of a particular institution, which have been identified by Pollack (1996, p. 434), Wendt (2001, pp. 1023–4) and Wallace (2000a, pp. 43–4) as rational expectations about costs and benefits (Pollack, 1996, p. 434; Wendt, 2001, p. 1043) as well as normative considerations of what is appropriate (Wendt, 2001, p. 1023–4). Héritier (2007, p. 1) on the contrary suggests institutions are chosen unintentionally. Also, it will be interesting to see in how far the founders of the institutions were constrained by the availability of institutional alternatives and borrowed from prior institutions, as suggested by Jupille (2004, p. 1) and Scott (2008, p. 94). Additionally, the question arises in how far

- contextual factors, derived from the broad circumstances;
- functional factors, derived from some of the core functions of politics and demands to be met, and
- institutional factors, i.e. opportunities and constraints which influence in how far institutions are able to channel and to structure the behaviour of political actors from the participating countries (Wallace, 2000b, pp. 43–4),

were relevant in the process of institutional selection in the case of ESRO, ELDO and ESA?

4.1.1 Phase 1: The Historic Evolution of ESA (1959–1974)

According to the theoretic framework identified, this phase should comprise the stage of agenda formation comprising the emergence of a topic on the political agenda, the framing of the issue for consideration in different forums and "the rise of the issue to a high enough place on the international agenda to the point of agreement on the provisions of a specific regime" (Levy et al., 1995, p. 282).

The case study has shown that instead, the first phase covers all five stages of institutional development, i.e. the agenda formation (the need to coordinate European space efforts through a European institution), the institutional choice (how ELDO, ESRO and ESA were designed), the "operationalisation" (of ELDO, ESRO), the redefinition of ELDO and ESRO and finally their crisis, making room for the foundation of ESA. This observation could be considered a proof of the permanent process of institutional development and the stages of institutional development as suggested.

Coming back to the conditions for an institution to be founded as identified based on Rittberger and Zangl (2006, p. 24), DiMaggio (1988, p. 14) and Olsen (1997, pp. 209–10, 217), i.e. the "problem condition", the "cognitive condition", the "hegemonic condition" (Rittberger & Zangl, 2006, p. 16), "windows of opportunity", and breaking points, the case study showed that all of them were present when ESRO and ELDO were founded. The need and wish to be able to compete with the U.S. and the will not to be left behind, can be considered as the main reasons, pushing European States to cooperate in the space field. This "problem condition" can be identified as the reason for the foundation of ESRO, ELDO and later ESA. The successful example of CERN, which gave rise to calls for collaboration in the area of space, provided for the "cognitive condition", i.e. the understanding that such scientific cooperation is best to be achieved through cooperation within a European organisation, which could pool the resources of the European countries. In line with Olsen (1997) the foundation of ESRO can also be considered as an example of institutional foundation through the exploitation of a window of opportunities. Both the success of CERN and the linkages between scientists established in the International Geophysical Year, helped Amaldi and Auger in their institution building effort. The force of this window of opportunity is particularly striking in the light of Amaldi's previous attempt to found a European space organisation modelled to CERN, which had resulted in warnings against a straightforward application of the CERN model to space (Krige & Russo, 2000, p. 18). The "hegemonic condition" can be said to have been present through the strong involvement of Britain, which was mainly provided by Massey. In line with DiMaggio (1988, p. 14), new institutions arise when organised actors with sufficient resources see in them an opportunity to realise their interests. In light of the formal cancellation of the Blue Streak programme, the British proposed a European research programme based on the use of Blue Streak as a launcher, as they saw this as their only opportunity left to realise their interests, i.e. a European launcher. The cancelation of the Blue Streak programme can, in line with the above-mentioned conditions, also be considered as a breaking point.

The case study under discussion also provided interesting insights into the relevant factors for institutional selection, -design and choice. Both ELDO and ESRO have been selected intentionally based on rational expectations about the costs and benefits expected through a process of intergovernmental bargaining on the basis of national preferences and other motivational factors. A European space institution as such was proposed by Auger and Amaldi and the scientific community. Considering CERN as a role model, they were initially torn between different institutional models such as the EURATOM or a European NASA, the latter not being acceptable to European States due to the military aspects. At the Meyrin Conference, States also started to question the CERN model as in contrast to CERN's particle accelerator, rocket and satellite technology is adaptable and can be used militarily. According to Sheehan (2007) the governments also consciously decided not to pursue the possibilities of ESRO being sponsored by the UN, or even by the Organization for European Economic Cooperation, because this would allow non-European participation (the U.S. and Canada were about to join the OEEC) and reduce its impact as a vehicle for European cooperation (Sheehan, 2007, p. 77). Thus, it can be concluded that institutional alternatives influenced the institutional choice of the founders of ESRO and ELDO and in line with Jupille (2004, p. 1) and Scott (2008, p. 94) borrowed from prior institutions.

The UK's refusal to discuss the launcher question can be seen as an example for intergovernmental negotiations to be dependent on the relative bargaining power of governments. UK had a great bargaining power due to its experience in the Blue Streak programme. As a consequence, a unified institution was no longer feasible. The result was an agreement on two space organisations, one similar to CERN, dedicated to basic research in space sciences and the other one to the development of launchers (Krige, 1992, pp. 1–2; Krige, 1993b, pp. 1–2). This is in line with Jupille's (2004, p. 1) suggestion about actors seeking to ensure the usage of institutions that maximise their political influence in the light of the available institutional alternatives.

At the same time, the example of ELDO and ESRO also gave insights into the question of institutional stability. While ESRO proved to be successful and was transferred and renamed as ESA, ELDO was struggling right from the start. One possible explanation to this discrepancy could be that ELDO was created based on political will, while ESRO was created given the growing recognition among certain communities, primarily the scientific community, of the stakes involved in space (Gaubert & Lebeau, 2009, p. 37).

4.1.2 Phase 2: From the Founding of ESA to the Entering of the EC in the Space Domain (1975–1986)

The second phase of the development of space collaboration in Europe was characterised by the early years of ESA. It covers the phases foundation, "operationalisation" and "consolidation"[2] of the stages of institutional development. While in the first phase the basis for the foundation of a new institution had to be laid including Levy et al.'s (1995) stage of "agenda formation", ESA could build upon the foundations of ESRO and ELDO with the objective to consolidate space collaboration in Europe in the second phase. It thus started directly with Levy et al.'s third phase "operationalisation". This development is in line with Florensa's (2004, pp. 9–10) and Dehousse and Magnette's (2006, p. 18) phases of institutional development referred to as "consolidation".

In line with Rittberger and Zangl (2006), the problem condition as well as the cognitive condition, which were present for the foundation of ESRO and ELDO, were also valid for the foundation of ESA. In contrast to ESRO and ELDO, which came about through the exploitation of a window of opportunity, ESA came into being through a grand decision (Olsen, 1997, pp. 217–9). It is not possible to identify one hegemon fulfilling Rittberger and Zangl's (2006, p. 16) "hegemonic condition". The blow up of Europa 2 in November 1971 and the U.S.' offer to European governments to participate in the post-Apollo programmes (Madders, 1997, p. 140) can be considered as breaking points.

By the time of the foundation of ESA, the geopolitical context had changed (Sheehan, 2007, p. 84). The neo-functionalist ideology of European integration had begun to gather considerable momentum and integration/collaboration had begun to be asserted as a value itself. Consequently, ESA became understood as "a tool for working towards a united Europe as it were a 'European Space Community'." (Ibid., p. 83) In line with this, Sheehan (2007) considered

> that the work of ESA is not in the execution of research and development in space science and technology, rather the work of ESA is cooperation. (…) ESA's policies are a functionalist effort to concretise the practice of unity in diversity that has characterised the European integration project since the early 1950s. (p. 83)

Sheehan (2007, p. 84) also refers to European space science and technology having significantly matured and with a decade of cooperative experience within the space organisations and within the EEC itself. Accordingly, the political views of the European States had also matured and Europe was now capable of acting as an entity with a clear political will and consistent policy. At the beginning of the 1960s, Europe was not ready for this yet, as can be seen in the decision to create two separate space organisations. While several institutional heritages such as ESRO's Convention were taken as a basis for ESA's institutional structure, ESA was also restructured to overcome the programmatic problems that had hampered ELDO and ESRO. The institutional choice was based on rational expectations about costs and benefits as well as functional and institutional factors. Bearing in mind the experiences the founders had with ELDO and ESRO, several institutional alternatives have also been considered already. As a consequence, the new ESA was created to consist of a two-tier structure of finance and involvement. ESA was thus right from the start on a better footing than ESRO and ELDO, which can be considered as a stabilising factor.

[2]For example with "with manned space programmes" in the 1980s (cf. Madders, 1997, p. 588).

Providing for the basis for the foundation of ESA, this phase is of particular interest to the question of the future of ESA, since knowing how institutions were constructed "provides insights into how they might come apart" (Thelen, 1999, p. 400).

4.1.3 Phase 3: From the First Steps of the EC to the ESA-EU Framework Agreement (1986–2004)

The third phase of the development of space collaboration in Europe has been characterised by the first steps of the EU's Commission in the 1990s to show interest in space technologies. ESA was no longer the only "space Europe". Rather cooperation had to be discussed and initial ad hoc arrangements agreed upon. This is why Eurospace, the European space industry association, refers to a turning point and "the first phase of ESA's activity" (Eurospace, 1995, p. 230), being "essentially successful and complete" (Ibid.). In so far the third phase differs from the previous two, as space activities were now divided into the framework of the "two Europes", constituted by the EU and ESA (Jourdain, 1995, pp. 87–8; Suzuki, 2003, p. 201). This development is in line with Florensa's (2004, pp. 9–10) and Dehousse and Magnette's (2006, p. 18) phases of institutional development referred to as redefinition, i.e. "relaunch" and "adjustment" or "redefinition". In contrast to the previous phase, ESA had to adjust to another space player.

4.1.4 Phase 4: From the Framework Agreement to the Lisbon Treaty (2004–2009)

The fourth phase of the development of space collaboration in Europe has been characterised by the formalisation of the EU-ESA cooperation through the ESA-EU FA and the subsequent introduction of the Space Council. Based on the previously agreed White Paper, the relations of both players entered a new chapter. At the same time, space applications became increasingly used in the area of security and environment. The Commission's continuous interest in space technologies resulted in space as an area of Research and Development, being first incorporated into the Draft Constitution and finally the Lisbon Treaty. After a phase of institutional adjustment and redefinition, as Florensa's (2004, pp. 9–10) and Dehousse and Magnette's (2006, p. 18) put it, this phase laid the foundation for the cooperation between EU and ESA. This phase can be referred to as "consolidation", given the formalisation of the previously ad hoc cooperation between EU and ESA. This proves that a redefinition does not necessarily have to result in a crisis but an institution can also consolidate instead.

This phase can be considered the era of the Space Councils. As will be seen the initial enthusiasm about the Space Councils has withered away and their significance continued to decline (Köpping Anthanasopoulos, 2016, p. 82). The repetitiveness of the resolutions particularly after 2009 shows the declining importance of the Space Council meetings, which also became significantly shorter. While the first one took about two hours, the last one took only mere 65 min (Ibid., p. 88).

4.1.5 Phase 5: The Lisbon Treaty: Introduction of Space as a Shared Competence (2009–2012)

The fifth phase too can be put under the heading "consolidation", being dominated by discussions on the impact of the Lisbon Treaty, the resulting shared competence in space of the EU and their impact on the FA or on a more general level on the future of the relationship of EU and ESA. This phase can also be considered the climax of the rapprochement of the EU and ESA with the Commission Communication entitled "Establishing appropriate relations between the EU and the European Space Agency" (European Commission, 2012) of 2012, being the first document to explicitly mention the governance question in its title. This phase shows the difficulties in applying the stages of institutional development given the difficulty in distinguishing between "consolidation" and "revision".

The dialectic of the discussion on space also changed during the fifth phase. Space applications were put in the light of the benefits for citizens in terms of providing answers to global challenges.

4.1.6 Phase 6: Post-ESA-EU Declaration of November 2012 (Post-2012)

While the discussion on the relation between EU and ESA has gathered pace in 2014 it also lost in topicality. Observers argue that the discussion remains largely characterised by vanities. The option of ESA becoming integrated into the EU seems to have faded, as the window of opportunity opened by the discussions on the Multiannual Financial Framework, which was agreed upon on 27th June 2013, is now closed. The sixth can thus also be referred to as "consolidation" or "revision".

4.2 Facets of Institutional Change and the Development of Space Policy in Europe

As has been explained, the existing literature offers surprisingly little discussion and guidance about how to determine institutional change empirically (Campbell, 2004, p. 5). To specify the most important aspects in describing institutional change

(Keohane & Nye, 2001, p. 50), different facets of institutional change have been identified based on the existing literature:

- the degree of change (incremental or revolutionary change; in which dimension it took place),
- the pace of institutional change (fast and slow changes),
- sources of institutional change (endogenous, exogenous),
- change agents and
- forms of institutional change.

In the following the case of ESA will be analysed in terms of these five facets.

4.2.1 Degree of Institutional Change in the Development of Space Policy in Europe

This dissertation, in line with Hall (1993, p. 278) and Campbell (2004, pp. 36, 48), has suggested to differentiate between

- simple change, i.e. in instruments and means,
- complex change, i.e. change in the normative (e.g. principles prescribing goals of behaviours and appropriate ways to pursue them) institutional dimension,
- change in the regulative (e.g. legal, constitutional, or other rules that constrain and regulate behaviour) dimensions of an institution, and
- change in the cultural cognitive dimension (e.g. culturally shaped assumptions about reality and the frames through which it is perceived, understood and given meaning) (Campbell, 2004, pp. 36, 48),

as to qualify the degree of change.

Applying this differentiation to the six historic phases under discussion, shows that the mere distinction between simple and complex change as suggested by Hall (1993, p. 278), is too simple to grasp the complex nature of institutional change. Namely all six phases have undergone simple change in instruments and means. Even the replacement of ELDO and ESRO through ESA can be considered to be a simple change, since ESA has the same goals and ends as its predecessors. Similarly, the following examples of the case study can all be considered changes in instruments and means, qualifying for simple change

- the introduction of new instruments such as the geo return principle, industrial policy as well as a manned space programme by ESA with the objective to increase the efficiency of the European space efforts, by making better use of the resources and to define a European space programme for exclusively peaceful purposes (phase 2),
- the introduction of different informal practices of cooperation of EU and ESA developing on a variety of occasions (phase 3),

- the formalisation of their collaboration through the ESA-EU FA, committing the two organisations to working together, through the introduction of the Space Council (phase 4),
- the introduction of the shared competences (phase 5),
- and the assessment process of EU–ESA relations (phase 6).

Thus, a differentiation between dimensions of institutions in which change took place, as suggested by Florensa (2004, p. 9), Hall (1993, p. 278) and Campbell (2004, pp. 36, 48), seems to offer a more complete picture. The founding of ESA with a new ESA Convention (phase 2) provided a new legal basis of the newly founded institution, which can be considered a change in the regulative dimension. Similarly, both the FA (phase 3) as well as the Lisbon Treaty (phase 4) can be considered to provide changes for ESA, regarding the regulation of its behaviour, and can thus be considered as changes in the regulative dimension. The mere interest of the EC in the space domain at the beginning of the third phase is neither a change in the regulative nor in the cultural dimension and thus cannot be grasped by the proposed theoretic framework. The same holds for the developments since 2011, when ESA started to struggle defining its role post the Lisbon Treaty. In practice, it also seems to be rather difficult to identify change in the cultural cognitive dimension, which does not seem to be an objective characteristic but can be rather subjective. Thus, there is still need for a different more distinguished theoretic account to qualify the degree of change.

4.2.2 *Pace of Institutional Change in the Development of Space Policy in Europe*

The second dimension suggested to qualify institutional change, was pace. Drawing on the literature at hand, it was proposed to distinguish between two types of paces of institutional change:

- slow, incremental through small adjustments, gradual or evolutionary in a continuous manner and
- rapid revolutionary change through large steps in a discontinuous manner, resulting in a "steep transformation", breakdown and replacement of old institutional arrangements (Campbell, 2004, p. 33; Madariaga, 2011, p. 17, 29; Roland, 2004, p. 116–7).

Similar to the findings regarding the degree of change and in line with many Historical Institutionalists (i.e. March & Olsen, 1989, p. 169; North, 1990, p. 73; Pollack, 1996, p. 441), who consider incremental change as typical, referring to it as the "main story" (March & Olsen, 1989, p. 169), four out of six phases under analysis can be considered to have changed in an incremental manner.

Only the founding of ESA (phase 1 and phase 2), replacing ELDO and ESRO, can be considered a large step in a discontinuous manner. It is further possible to distinguish the pace of institutional change, taking place at the two institutions ELDO and

ESRO. The pace of institutional change of ESRO qualifies as slow and incremental, while ELDO's pace of institutional change can be considered revolutionary. The main difference of both institutional developments is that in the case of ESRO, its Council had already set up a group of experts to study the internal structure, which had proposed a revised internal structure already in 1966 (Krige, 1993a, p. 37). ELDO however, regardless the foundation of the ESC with the objective to gradually harmonise European space activities, had taken no steps to revise until the Member States' decision in 1973 to transfer ELDO's staff to ESRO. The latter step can thus be considered revolutionary in nature.

The institutional changes in the following phases were of incremental nature. Neither the introduction of different informal practices of cooperation of EU and ESA, developing on a variety of occasions (phase 3), nor the formalisation of their collaboration through the ESA-EU FA, committing the two organisations to working together through the introduction of the Space Council, can be considered large steps in a revolutionary manner.

The incremental changes taking place during the previous phase seem to have culminated, resulting in an evolutionary change with space being incorporated in the Lisbon Treaty as a shared competence of EU and Member States. This can be seen as a result of the continuous push of the Commission to gain competencies in this field. It can be considered a prime example for small incremental steps accumulating to a revolutionary change. While previously there were mainly political declarations in form of Commission Communications about the EU's need to become a space player, there was now a legal basis for the EU to act in the space policy field, which went further than the two flagship programmes Galileo and Copernicus.

4.2.3 Sources of Institutional Change in the Development of Space Policy in Europe

It was further proposed to distinguish between endogenous sources of institutional change (cf. Greif & Laitin, 1994; Streeck & Thelen, 2005; Young, 2010), such as

- self-undermining processes (Greif & Laitin, 1994, p. 634),
- institutional design (Héritier, 2007; Streeck & Thelen, 2005, p. 19) or internal contradictions (Young, 2010, p. 174),
- ideological innovation, learning processes through the acquisition of knowledge and skills and changes in tastes or preferences of the actors involved (Florensa, 2004, p. 6),

and exogenous sources of institutional change (cf. Krasner, 1988, p. 67; Pollack, 1996, p. 441; Scott, 2008; Steinmo, Thelen, & Longstreth, 1992; Suzuki, 2003), such as

- "critical junctures or punctuation" (cf. Historical Institutionalism), i.e. changes in the attribution of natural resources, extreme climatic events, or natural disasters,[3]
- changing preferences of supportive/influencing governments (Pollack, 1996, p. 441), which might result from unfulfilled expectations,
- shifts in power distributions of previously supportive coalitions or the appearance of new actors (Ibid., p. 438; Florensa, 2004, p. 18; Scott, 2008, p. 198),
- changing voter preferences, leading to new political alignments,
- lobby of business interest (Scott, 2008, p. 196),
- broad changes in the policy environment of social, political, or economic nature (Florensa, 2004, p. 18; Pollack, 1996, p. 438),[4] and
- improvements in the quality of information about either the operation of the institution or the environment within which it operates (Pollack, 1996, p. 438).

In four out of six phases under analysis both endogenous and exogenous sources of institutional change could be identified. Since the second phase mainly dealt with the "operationalisation" and founding of ESA, naturally neither endogenous nor exogenous sources for change could be identified. Additionally, the differentiation between endogenous and exogenous sources of change is difficult during the fourth phase as this is characterised by the resolutions issued by the Space Council consisting of both the ESA Council and the EU Competitiveness Council. The two main exogenous sources which could be identified in the six phases under analysis were the impact from the European integration process and from the geopolitical context mainly coming from the U.S. In five out of six phases under discussion, institutional change can be said to have resulted from the geopolitical context. This is in line with Steinmo, Thelen, & Longstreth's, (1992) model of "institutional dynamics", which focuses on the interaction between institutions and political processes. Accordingly, changes in institutional contexts of socio-economic or political nature can influence existing institutions.

Endogenous and exogenous sources for institutional change as identified will be further distinguished and exemplified below.

4.2.3.1 Endogenous Sources

In the first phase, it is possible to identify several endogenous sources of institutional change, which were enshrined in the institutional design of both ESRO and ELDO. While ESRO is largely perceived as having been relatively successful in fulfilling the objective it was set up for, i.e. to encourage European collaboration in space research and technology, its demise can largely be attributed to three observations: (1) the relative youth, inexperience and euphoria of the founding community, (2) the

[3]For more information on "critical junctures" refer to the following literature (cf. Campbell, 2004, p. 26; Christiansen & Vanhoonacker, 2008; Florensa, 2004, p. 18; Hall & Taylor, 1996, p. 942; Horak, 2007, p. 21; March & Olsen, 2006, p. 12).

[4]For example changes in the markets, technological innovations or legislative changes (cf. Steinmo et al., 1992, p. 16).

avoidance of competition with national space programmes, which limited ESRO's expenditures, (3) ESRO being only one among several options for scientists given NASA's bilateral agreements to fly scientific satellites (Krige, 1993a, pp. 26–7). Among the endogenous sources for institutional change of ESRO were thus: (1) the hopelessly inadequate financial resources, especially in the light of the high administrative expenses, (2) the malfunctioning established structure, resulting in an agenda crowded by relative trivialities, (3) a self-undermining process due to a "crisis of confidence" resulting from an unclear distinction between legislative and executive parts of the organisation (Ibid., p. 38). ELDO was also subject to institutional change due to endogenous sources. Among them were (1) an inadequate budget due to inexperience, (2) difficulties compounded by the policy of "juste retour" inherent in the Convention, which could not be achieved through the Europa programme, since the main elements had been allocated to a limited number of countries for reasons of efficiency (Sheehan, 2007, p. 78) and (3) the lack of an institutional mechanism neither for political discussion within the organisation, nor for coordinating the activities of ELDO and ESRO (Ibid.). This institutional separation of ESRO and ELDO proved to be impractical and rather hampered the deployment of competitive space activities. Thus, by 1970 ELDO had neither managed to launch a single satellite with the Europa rocket, nor had it achieved anything which would have been beyond the financial and technical capabilities of any of the larger Member States (Ibid., p. 81) and thus did not fulfil the expectations it had been founded for.

Since the second phase mainly dealt with the "operationalisation" and founding of ESA, naturally no endogenous sources for institutional change could be identified.

The third phase was also subject to endogenous sources of institutional change. One example of such a process is ESA revision process, which was started by its Ministerial Council as well as several working groups established by ESA's Director General (DG). It questioned ESA's mission and evaluated its role vis-à-vis other institutions. The result was a call for a more coordinated approach among the European space actors, as well as a formalised resolution on the future of Europe in space (cf. Creola, 2001; ESA, 1999, pp. 20–5), asking the ESA Executive to prepare a fully developed strategy together with the European Commission to be delivered by the end of 2000. Additionally, three Wise Men were asked for an independent advice on the evolution of the ESA-EU relationship, which resulted in a major turning point in the previously sectoral rapprochement of ESA and the Commission. The "Agenda 2007" (cf. Battrick, 2003) issued by DG Dordain in 2003, can be considered as proof of a new era. This endogenous process however can be considered a natural step in the institutional development stage of "institutional adjustment".

The differentiation between endogenous and exogenous sources of change is difficult during the fourth phase as this phase is characterised by the resolutions issued by the Space Council consisting of both the ESA Council and the EU Competitiveness Council. Based on the resolution issued by the Fourth Space Council, ESA's DG Dordain issued a so-called "ESA Director General's Proposal for the European Space Policy", which foresaw for the EU to take the lead in identifying and bringing together user needs and for ESA to develop technologies and systems (ESA, 2007). Building upon the previously mentioned Agenda 2007 of 2003, ESA's DG issued Agenda

2011 in 2007 with the objective to adapt "ESA to a changing environment". Therein he foresaw a shift of ESA's focus towards promoting space applications in areas such as security, disaster relief or maritime surveillance, thereby adopting a user-centred approach. Both documents can be considered examples of an endogenous revision process and institutional adjustments. Endogenously ESA's DG was active in the governance debate, adjusting the institution.

During the fifth phase, ESA's Member States discussed the future of ESA as a response to the inclusion of space as a shared competence of EU and Member States in the Lisbon Treaty. Among these, particularly France was pushing for ESA to be incorporated into the EU, while Germany made efforts for it to remain an intergovernmental institution outside the community framework. The role of ESA's DG Dordain, a French national, is particularly interesting. Instead of trying to keep "his" institution outside the community framework, he was, in line with the position of the French government, pushing for ESA to be incorporated into the EU. This can be explained by ESA being a strong competitor to the French national space sector. At the same time the head of the German space agency Wörner made strong efforts for ESA to be kept as it is. The political declaration entitled "Towards the European Space Agency that best serves Europe" (ESA, 2012), which was agreed upon by ESA's Member States in 2012 and aims at setting the framework for any deliberations about the future of ESA within Europe, can be considered a form institutional adjustment. It is a natural step to endogenously reconsider its own role, when a coupled institution enters the own field.

During the sixth phase, the Commission tasked the independent consultancy Roland Berger to provide a cost-benefit analysis of the options. As a consequence the ESA Executive commissioned a study as well. It consisted of a scenario assessment. Again, the role of ESA DG Dordain can be highlighted during this phase.

4.2.3.2 Exogenous Sources

The first phase highlights how broad changes in the policy environment can affect institutions as an external source of institutional change. At the same time it also shows how changing preferences of supportive governments can lead to institutional change. In this context, a decisive role can be attributed to the impact of the European Integration process as well as the general geopolitical context, being characterised by the super power rivalry. Although space developments have been independent of the general process of European integration in Europe, ESRO and ELDO did not operate in a vacuum. For example, the negotiations over the setting up of ELDO took place in the context of the UK's growing wish to become part of the EEC. Offering leadership in the field of European space cooperation was "one dimension of a wider strategy aimed at closer integration" (Sheehan, 2007, p. 78) and was seen as an important proof of Britain's (new) European credentials (Ibid.). This perspective was echoed by West Germany, whose government welcomed the British proposal for a European launcher organisation as "an opportunity to strengthen the linkages of the UK with the continent as such and also as a possible first step towards and enlargement of the

EEC" (Ibid.). Consequently, ELDO was caught up in the internal political tensions, such as the political opposition between France and the UK on basic international political strategy[5] and the British EEC membership. The critical interrelationship between European space policy and the wider politics of European integration can be exemplified by the French veto on Britain's application for EEC membership in November 1967, which resulted in the reduction of the British contribution to ELDO in favour of raising funds for a national space programme in 1966. The French decision can be considered an example of a critical juncture.

Additionally, during the first phase, the external environment of the "space race"-era shaped and limited the choices of institutional arrangements at European level (Suzuki, 2003, p. 80). As Krige (1994, p. 2) underlines, the European space system was not a closed but a dynamic and unstable system, subject to external shocks. Many observers argue the main impetus for European States to collaborate was to compete with the U.S. (Suzuki, 2003, pp. 2–3). A particular example is the post-Apollo programme, which necessitated a comprehensive space programme on the European side, from launch pads to launchers, satellite technology and the capability to maintain humans in space. Additionally, there was a case for independent access to space for Europe (Hörber, 2016b, p. 16). Also, after a few bad experiences with NASA resulting from frequent changes in U.S. policies on technology sharing and trade, many European States preferred to work in a reliable European space organisation rather than remaining dependent on assistance from the U.S. (Hörber, 2016a, p. 58). Apart from the fear to be left behind by the U.S., it was also the policies of the U.S. government and the resulting commercialisation of the space market that impacted upon the developments in Europe (Suzuki, 2003, p. 1). When in 1965 the U.S. satellite Early Bird demonstrated for the first time the feasibility of telecommunications by satellite, thereby "opening up an entire new field of immense commercial and military potential" (Krige, 1994, p. 2), the objectives of the European programme were re-evaluated. The commercial success of telecommunication programmes by the U.S. government and industry changed the policy environment. As a consequence, ELDO's approach was no longer considered sustainable. In line with Historical Institutionalism, which expects changes to occur only under extraordinary circumstances, i.e. when changes in the external environment occur, this event can be considered as a critical juncture, leading to a new path, namely the foundation of ESA. In line with Historical Institutionalism extraordinary circumstances mark a departure from the existing "path" (Christiansen & Vanhoonacker, 2008).

In contrast to the first phase, discussions in the context of the wider politics of European integration did not have an impact on the space field during the second phase. This can be explained by Europe's integration ambitions coming to a halt in the 1960s and 1970s within the EC, when the Franco German differences caused a stalemate and put a stop to the efforts to achieve a common design for nuclear reactors (Borras, 2006, p. 1). It is thus particularly striking that integration in the field

[5]France wanted to achieve autonomy for Europe, and to avoid dependence on NASA and the U.S., while Britain was opposed to any initiative that might weaken the links between the U.S. and ties European allies.

of space still proceeded. The geopolitical context outside the EU also influenced the space developments in Europe during the second phase. In contrast to the previous phase, it was no longer the fear to be left behind the U.S., but mainly the wish to be autonomous that gave the impetus to further collaborate. ESA's Convention has also been influenced by the headway of the U.S. and the USSR, leading to the inclusion of industrial policy as a purpose of ESA (Suzuki, 2003, p. 87). ESA's Member States aimed to manifest the independence of European capabilities in space, which was driven by political and institutional reasons based on the experiences with the Aerosat programme.

When the European Community gained competencies in the area of research and development through the SEA and space applications became increasingly important for the policy areas of the EC during the third phase, the EC's resulting gradual development as a space player impacted on ESA's development. The most obvious example is the need for an agreement on the division of labour between EC and ESA, which resulted in the FA. The sudden interest of the EU to engage in the field, is often explained by relying on Neofunctionalism and its "spill-over"- thesis. In line with this, Cheli and Schrogl (1999, p. 63) highlight the integration of space technologies and applications in daily life as the main impetus for the European Commission to become interested in space technology. Space applications generally came to affect and interest a wider range of actors and users than in the past. As time passed, the use of space technologies became an increasingly important tool for the achievement of a considerable amount of policy aims in areas in which the EC/EU possessed competences (such as for instance transport infrastructure, environmental protection, and the information society) and thus an interest in space policy evolved (Hobe, Kunzmann, & Reuter, 2006, p. 537; Peter, 2007, p. 101; Reuter, 2007, p. 12). Space had become an important instrument for the implementation of many of the EU's policies (Hobe et al., 2006, p. 540). The Commission for example became principle customer of the Earth Observation Data among other things in the context of the Common Agricultural Policy (Reuter, 2007, p. 12). Moreover space was understood to contribute as a symbolic feature of European integration representing the EC on the world stage (Suzuki, 2003, p. 185). Establishing a European space policy was considered a means for the EC to be a better neighbour and respected partner of the global community (Hesse & Hornung, 2012, p. 185), to improve its reputation in the world, to strengthen economic competition as well as to increase its reputation in research and development (Ibid.). The introduction of Galileo and Copernicus, the two flagship programmes of the EU's Commission, further highlighted questions related to governance and ESA-EU relations during the third phase. Space was "going through a period of major re-evaluation and change" (Eurospace, 1995, p. 227), as it was "no longer driven by the political and military rivalries of the Cold War" (Ibid.) and governments were "reviewing their space aspirations and budgets." (Ibid.). For a long time, European and national space agencies had played a central role in defining space policies and implementing them with considerably large budgets and relative autonomy. Competition and budget pressures resulted in changing roles and importance of national space agencies. The demand for space programmes had become increasingly user-and commercial-oriented. Space agencies were no longer able to guarantee the best programmes and highest salaries given the limited financial resources (Suzuki, 2003, p. 145). At the same time, space applications became increasingly used in the area of security and environment. The foundation of the GSA further adds to this complex relationship and competes with ESA for the status of the European space agency.

During the third phase, it was the wish to be independent and autonomous from the U.S. that led to changes in the context of satellite navigation, which had an impact on the developments in Europe. European airlines as well as other users of satellite navigation services relied heavily on the U.S. GPS, which, completed by 1993, was controlled by the military. It was the U.S. Department of Defense's option for discontinuing civilian usage in case national security, which was the case during the Kosovo crisis, resulting in disturbed air traffic in Greece and Italy, that led to the European Commission's DG Transport's idea of developing a European system as an alternative.[6]

The review of the legal framework of the EU provided a window of opportunity to introduce space as a community policy area during the fourth und the fifth phase. The fact that space is explicitly referred to in the Lisbon Treaty reflects the growing recognition of space systems and related technologies, providing beneficial products and services to European citizens (Times Higher Education, 2004). The European integration process, i.e. the introduction of space as a shared competence by the Lisbon Treaty, also influenced ESA. The continuous push of the Commission to include space policy as a community policy field finally gained ground. The geopolitical context also worked in favour of the Commission. In the context of the economic crisis, it was easy to frame space policy as an asset in the fight against the global challenges. The resolution of the sixth Space Council entitled "The Contribution of space to innovation and competitiveness in the context of the European Economic Recovery Plan, and further steps" (cf. Council of the European Union, 2009) exemplifies this. Among the exogenous sources for institutional change is also the economic crisis.

The geopolitical context and particularly the U.S. also influenced the development of space policy during the fourth and fifth phase. "[O]ne of the catalysts" (Suzuki, 2003, p. 193) for the foundation of the Space Council "was the question of European dependence on the U.S. satellite navigation system" GPS (Ibid.). It also gave the impetus for the EU to provide the framework for the European satellite navigation system Galileo. The Galileo programme was undoubtedly a new step in European space collaboration, driven by economic motivations as well as reasons related to autonomy (Ibid.). At the same time, as described by Dupas, Janischewski, von Kries, and Schrogl (2001, p. 103), European industry merged in an attempt to enable European companies to counterbalance the dominant U.S. corporations. The result was a changing space landscape in Europe in 2001. Additionally, traditional cooperation in space with international partners was replaced by competition. To face this competition, it became increasingly important for Europe to act unified (Cheli & Schrogl, 1999, p. 63). In the light of the new U.S. space policy, as a consequence of Obama winning the election and China becoming an established member of the space com-

[6]This section is based on Suzuki (2003), pp. 192–3.

munity, the competition in research and development was rising, calling for Europe to get its act together. At the same time, the world was ever more dependent on space applications, giving greater importance to space policy, especially with an increasing utilisation of space applications also in the field of security.[7]

The strong push from the Commission continued in phase 6. No further influence of the geopolitical context for example coming from the U.S., could be identified in phase 6.

4.2.4 Change Agents in the Development of Space Policy in Europe

In line with Madariaga (2011) it is proposed that sources of institutional change "have to be activated by agents willing to generate change" (Madariaga, 2011, p. 24), such as States (and their ministries, representatives), individuals, groups (interest groups and communities of experts) or institutions (administrative staff of international organisations; parliamentary assemblies of international organisation) (Rittberger & Zangl, 2006, p. 78).

In all six phases under discussion, States acting as change agents could be identified. In five out of six phases, individuals could be identified to have acted as change agents. In two out of six phases, individuals who were part of an epistemic community, acted as change agents and are thus considered as groups acting as change agents. In four out of six phases, the Commission, an institution, can be identified to have acted as a change agent.

Change agents as identified will be further discussed and exemplified for the case of space developments in Europe below.

4.2.4.1 States

During the first phase, space policy was conducted exclusively by individual States, which thus were the main change agents. While the agenda formation can be traced back to Auger and Amaldi and the scientific community, the question related to the institutional selection, -choice or -design was made by States, leading to the decision to found not one but two institutions dealing with space policy. One of the main objectives of European States at the time was to keep up with the U.S. This is in line with Neorealism, explaining the evolution of European space collaboration as a consequence of the convergence of national interests (Suzuki, 2003, p. 3) and a response to the dynamics of superpower rivalry (Rosamond, 2000, p. 133). Changes in European space collaboration are understood as State-oriented and self-interested activities. Accordingly, all governments participating pursue it (1) to gain economies of scale;

[7]For further information on Europe as an actor in space and security (cf. Rathgeber & Remuss, 2009; Remuss, 2010, 2011).

(2) to avoid or share research and development risks; (3) to increase financial support from partner governments; (4) to increase the market size for the product; and (5) to strengthen the European industry to win and survive in the international competition with much larger U.S. counter parts. European space collaboration is presented by Neorealists as the "coalition of the weaks" with the preferences of European States converging in the light of the big difference of technological capabilities compared to the U.S., arguing that the main impetus for European States to collaborate was to challenge the U.S. (Suzuki, 2003, pp. 2–3). Governments were determined to maintain control over the form of collaboration in light of their national interests and considered different institutional designs such as CERN, EURATOM, a European NASA or an UN or OEEC funded institution (Krige, 1993b, pp. 1–2). The discussions at the Meyrin Conference are an example for the intergovernmental bargaining over institutional design, as suggested by Pollack (1996, p. 434), Wendt (2001, p. 1043) and Koremenos, Lipson, and Snidal (2001, p. 762). Proponents of Rational Choice Institutionalism argue that European countries collaborate because they want to reduce risks and cost of research and development, thereby strengthening national industry and increasing financial support (Suzuki, 2003, p. 10). Britain for example initially sought to cooperate to rescue their obsolete Blue Streak Programme. European space cooperation was thus a means to achieve a national objective. This is in line with Suzuki's (2003, p. 3) assumption of cooperation being the "second best option", compromising both the security of jobs, technological spin-offs, export potential and funding problems of capital-intensive space programmes. Hörber (2016b, pp. 24–25) refers to the British motivation as "Britain's budget frame", arguing that Britain was in favour of a European space programme to get value for money. Accordingly, France was driven by an independence frame, aiming at developing a space industry that would be competitive in innovative technologies.

The French veto on Britain's application for EEC membership in November 1967 exemplifies Rational Choice Institutionalism's and Neorealism's assumption of institutional change or replacement due to preference shifts of supportive governments (Héritier, 2007, p. 12; Suzuki, 2003, p. 2). Moreover, institutional change can also be considered as resulting from unfulfilled expectations. As ELDO's budgetary and technological problems hampered it, in fulfilling the objectives, it was designed for and led to a reassessment of Member States' commitments to the programme. Because of their frustration about the performance and industrial and technological return, the British industrialists for example published a statement in May 1966 urging the government to move away from the priority of the launcher programme within ELDO, towards a national one resulting in a reduction of the British financial contribution (Suzuki, 2003, p. 2). Also, both ELDO and ESRO experienced reconsiderations of the commitment of the smaller countries,[8] who were frustrated that their industries were not benefiting from the membership in relation to their contribution. In 1967 Spain, for example, threatened to withdraw from ESRO if their industrial return was not guaranteed (Ibid., p. 63). As has been shown, this problem was solved through the introduction of the geo-return principle at a later stage. This can be considered

[8] for ELDO: Italy, for ESRO: Spain.

another example of institutional change due to unfulfilled expectations and Member States' reconsideration of their commitment.

Moreover, the powerful States France, the UK and Germany together with Italy created the institution ELDO, which disproportionately benefitted them, as they were trying to make use of their existing technological programmes and capabilities. When however, it turned out that their technological capabilities neither matched nor where possible to be combined with the budget allocated, their solution to the collective action problem became unstable. This is in line with Neorealism, which holds that effective and long-standing cooperation amongst States, is impossible because of the essentially competitive and rational nature of the inter-State game. Neorealism views interstate cooperation as driven by the preferences of the most powerful States in that game (Rosamond, 2000, p. 133), which Waltz (1979, p. 199) calls the "managers". In this vein, the UK can certainly be perceived as one of the managers.

ESRO and ELDO can however also be looked at separately. When looking at ESRO, all States participated on the same basis: making sure that Europe was not left behind in the space field. ELDO however was created by States agreeing on the lowest-common-denominator, with the objective to realise different gains from cooperating in launcher development: The UK wanted to fund the Blue Streak, which it could not do on its own. France wanted to be independent of the U.S. and brought in its industrial know-how, and Germany hoped to establish "a capable industrial and technological base for the aerospace sector" (Suzuki, 2003, pp. 49–50), which would contribute to the integration of Europe and Germany, thereby bringing technological expertise to Germany (Ibid., p. 50). This is in line with Koremenos et al.'s (2001, p. 762) assumption about States choosing international organisation to fulfil certain self-interests and to further their own goals. Suzuki (2003) agrees, highlighting that ELDO was hampered from the outset by the fact that the different participating governments were driven by significantly different policy logics, economic for Britain, political for France and technological for Germany.

In addition to ESA's Convention being influenced by the progress of the U.S. and the USSR, ESA's Member States can be said to have been the only change agent during the second phase. Since its foundation, ESA is governed by the Ministerial Council. Therein each Member State is presented and has one vote. ESA's Ministerial Council provides the basic policy guidelines. Thus, it comes of no surprise that States are the only change agents. Nonetheless, the Ministerial Council held at Rome in January 1985 is often said to have marked a turning point for European space collaboration (Suzuki, 2003, p. 111), given its manifestation of autonomy and independence of European capabilities in space (ESA, 1985), which was driven by political and institutional reasons based on the experiences with the Aerosat programme. The introduction of the principle of juste retour and optional and mandatory programmes in ESA's Convention can be explained by relying on Neo-Realist arguments. Accordingly, States cooperate only because it is their "second best option", guaranteeing jobs in the national industry. The inclusion of optional programmes can be explained by the Neo-Realists' argument about the increasing vulnerability of a country's security, resulting from cooperation. Thus, cooperation is limited (Suzuki, 2003, p. 3). The introduction of industrial policy as a purpose of ESA and the principle of juste retour are today considered as the two main advantages of ESA as an intergovernmental institution outside the EU framework.

The third phase also provides an example for States' influence on institutions. As in the previous phase, it was Britain, who during the British EU Presidency in the first half of 1998 was able to mediate between the Member States and national space agencies, resulting in a joint resolution of ESA and the EU Competitive Council, which is often seen as a turning point, demonstrating the willingness of the Member States of both EU and ESA to strengthen the institutional leadership. As a consequence of the composition of the Space Council being composed of representatives of the Member States, States naturally were able to influence the debate on the roles of EU and ESA as part of the discussions within the Space Council.

The fourth phase is characterised by the resolutions issued by the Space Council consisting of both the ESA Council and the EU Competitiveness Council and thus the Member States of EU and ESA.

The fifth phase exemplifies that institutional change can also be the result of changing expectations of one of the Member States, leading it to reconsider its commitment. Many observers highlight that France increasingly perceived ESA and the European space programme as a competitor to its national space programme, thus favouring an increased role of the EU's Commission in the space field, knowing that it lacks experience and would thus not be able to compete with its national space programme. France was thus pushing for an incorporation of ESA into the EU.

Germany in contrast took the opposite position, by not supporting the existing efforts to integrate ESA into the EU, but considering an intergovernmental ESA to be necessary for a sustainable way of working (Wörner, 2013). A special role in the process of achieving the political declaration of the ESA Member States can be attributed to Germany with the German delegation making "a special effort in drafting" (Kraft, 2012, p. 17) the political declaration. The German position becomes very clear also in its dialectic: In Germany, it was discussed what role the EU was to play alongside ESA in the space sector and not the other way around (cf. Ibid). History has also shown that ESA can only make progress when France and Germany agree to compromise (Ibid.).

While favouring different scenarios, Member States of EU and ESA compromised in 2014 on keeping ESA as an intergovernmental organisation as laid down in the ESA Ministerial Council Conclusions of 2014.

4.2.4.2 Individuals

As has been highlighted, the International Geophysical Year together with the successful scientific collaboration of CERN during the first phase provided a window of opportunity, which Auger and Amaldi were able to exploit. The foundation of ESRO and ELDO would not have been possible without the engagement of Auger and Amaldi, acting as mediators between the different States. Additionally, the enthusiasm of Massey and the eagerness of Thorneycroft (Madders, 1997, p. 21) further added to the force of the impulse (Suzuki, 2003, p. 42). The first to call for the "creation of a European centre for rocket research" managed by scientists on the model of CERN was Blamont. Moreover, other personalities in charge of ESRO's and ELDO's management influenced the course of institutional development (Ibid., p. 81) through a series of choices adjusting them to a new environment (Ibid., p. 1). Those were Causse,[9] Bannier,[10] Spaey,[11] Puppi,[12] Bondi[13] and Aubinière,[14] who mobilised their ideas and efforts to find a middle ground of diverging national policies under certain institutional arrangements (Suzuki, 2003, p. 78). The case of European space collaboration thus shows that individuals can play an important, if not essential role in the agenda formation by highlighting the need for cooperation. As has been explained before Fligstein (1997a, 1997b, 2010), North (1990, p. 73) and Sabatier and Jenkins-Smith (1996) (Dudley & Richardson, 1999, p. 227; Sabatier, 1998, p. 109) have considered so-called "institutional entrepreneurs" as the main source for institutional change. According to Fligstein (1997a, 1997b, p. 17), the degree to which individuals are successful in acting as a change agent is dependent on their ability to frame "stories" and their "social capital", i.e. their acceptance and credibility in the community. This could explain that Amaldi was not successful in conveying his idea of Eurolune until he met Auger. Auger and Amaldi were considered as credible in the policy field, given their experience in founding CERN. Their social capital as well as their success in framing the story about the need for European space collaboration can be attributed to their past in CERN. In line with this Campbell (2004) has particularly highlighted the role of individuals as change agents in the context of bricolage; i.e. institutional change involving the recombination of old institutional elements (Campbell, 2004, p. 28; Lanzara, 1998, p. 26). Indeed, Auger and Amaldi, given their experience with CERN, were responsible for recombining institutional elements in innovative ways, leading to the institution building and the foundation of ESRO. Looking at the institution building process of ESRO, ELDO can be said to have lacked the "norm entrepreneurs" Amaldi, Auger and Massey (Suzuki, 2003, p. 52). On the contrary ELDO was the result of intergovernmental

---

[9]Jean-Pierre Causse was the Director of Brétany Space Centre of CNES, the French space agency and was entrusted with the task of drafting a report on the options for European space collaboration. His subsequent Causse Report, submitted in December 1967, contained several ideas on how Europe should manage space programmes and among other things, proposed for the first time to incorporate ESRO and ELDO into one organisation (Cf. 3.1.1.5).

[10]The Dutch civil servant Bannier, added a new programme structure to ESRO on the basis of the Causse report, designed to balance the need for solidarity and flexibility.

[11]In October 1968 Dr. Spaey formed the so-called Spaey group on the basis of Bannier's propositions and urged the creation of a single organisation.

[12]In 1971, the ESRO Secretariat set up a Steering Committee who, led by Puppi, issued the so-called Puppi Report, proposing Member States' participation on a case-by-case basis.

[13]Herman Bondi was ESRO's director in the mid-1960s and managed to find a compromise between smaller and larger Member States concerning contribution, benefit and international competitiveness.

[14]After the disappointing first flight of Europa 2, the ELDO Council asked Robert Aubinière, incoming ELDO Secretary General, to report on the causes of the failure. His subsequent report of June 1972 pointed to the lack of technical authority in ELDO's Secretariat (Cf. Suzuki, 2003, p. 78).

negotiations and not of scientific considerations. In line with Liberal Intergovernmentalism (cf. Moravcsik, 1993a) this was the lowest common denominator that the cooperating States could agree upon. According to Krige (1993c, p. 21), many experts in participating countries were sceptic, opposing ELDO, based on fears that the expensive rocket would be financed at the expense of their research programmes. Engineers stressed the obsolescence of the technology in the first stage, and the complex managerial problems that would be created by building bits and pieces of the system in different countries. These expert opinions were overruled in France, Germany, and Italy, along with the counter-suggestion that if Europe wanted to enter space rapidly, it would be advised first to negotiate to build a heavy U.S. launcher under license. Politicians and experts who favoured the founding of ELDO were also strongly encouraged by important sectors of the European aerospace and electronics industries. Consequently, ELDO was a child of political, not technical will (Ibid., p. 32). For ESRO the recognition among the scientific community for what was at stake proved decisive (Gaubert & Lebeau, 2009, p. 37). Thus, while Auger and Amaldi, the scientists, who are responsible for the agenda formation, initially wanted to set up an institution dedicated to both the scientific research as well as the joint development of a launcher, the construction of satellites and the scientific instrumentation needed for research, governments, and industry had different ideas about the nature and purpose of a future European body (Krige, 1992, pp. 1–2). While the scientists considered the launcher from a scientific perspective simply as a "means to put a scientific experiment into orbit" (Ibid.), for the politicians and industry representatives the launcher "was a device whose development was intertwined with national and European political, military and commercial strategy" (Ibid.). Given these two different perspectives on the matter, having initially thought to create a single organisation dedicated both to launcher development and to space research, Auger and Amaldi limited their ambitions to formulating a satellite and sounding rocket programme. This is in line with Florensa (2004, p. 8). Accordingly, individuals can realise a tension between their expectations and the conditions (of the internal and external environment) they face. Such asymmetries can lead to a situation where individuals perceive that their future expectations and interests could not be fulfilled, or the costs (or risk) associated are too high. In contrast, the scientific community retained fundamental control of ESRO. As a result, the scientists' low political profile provided a measure of independence and "shelter in difficult times" for ESRO. Moreover, the scientists remained at the centre of the organisation and understood what they were doing and what ESRO was doing. This was not the case for ELDO (Madders, 1997, p. 35).

During the second phase, no proofs of individuals acting as change agents could be identified.

The third phase of European space collaboration has also been influenced by individuals. Among these is ESA DG Rodotà, who tasked and identified the Wise Men, whose work led to the formulation of the European Space Policy. Rodotà also created four expert study groups, advising him on (1) the overall evolution of ESA, (2) the rationalisation of resources in Europe, (3) the relationship with the EU and synergies with other sectors and (4) relationship with industry and harmonisation of technology. At the same time, Lord Sainsbury, British Minister for Space, was able to bridge Member States' space agencies and ministries in charge during the British EU Presidency during the first half of 1998. Moreover, Delors' integrationist agenda (Borras, 2006, p. 1) led to the EC's entrance in the space arena. As has been explained the Space Advisory Group (SAG), which acted as the linchpin between the ESA Council and the EU Council of Ministers, was depended on the ideas of its Chairman. Consequently, SAG was more influential under the chairmanship of Jean-Pierre Contzen (1992–1996), an enthusiast for using SAG as a political framework, than under the chairmanship of Herbert Allgeier, who held a minimalist view on the Commission's role in space (Suzuki, 2003, p. 190). Contzen was also influential in 1992 as "space coordinator", by facilitating inter-DG discussion and formulating a coherent space policy for the Commission. As has been mentioned, he was also responsible for organising the Space Coordination Group (SCG).

The fourth phase of European space collaboration has also been influenced by individuals. For example, ESA's DG Dordain pushed the debate, by formulating governance ideas as published in the so-called "ESA Director General's Proposal for the European Space Policy", which foresees for the EU to take "the lead in identifying and bringing together user needs and to aggregate the political will" (ESA, 2007, p. 31) and ESA to "develop technologies and systems" (Ibid.).

In the fifth phase, ESA's DG Dordain has again influenced the direction of the discussion by first bringing the topic on the agenda, secondly repeatedly discussing it with the Commission and third providing the first draft for a declaration on ESA-EU relations. ESA's DG Dordain, a French national, did not act neutral but rather was playing the French card. As the head of ESA, one would expect him to support the survival of ESA as an independent institution. Instead, he was increasingly pushing for an analysis of governance structures, pointing to the duplication of efforts of ESA and the EU's Commission. Additionally, Johann-Dietrich Wörner, by then head of the German space agency DLR and chairman of the German delegation at the Ministerial Council Conference of 2012, has strongly influenced the agreed declaration by drafting the proposal himself, trying to find a compromise between the different Member States' positions. Wörner however can be said to have acted on behalf of his Minister.

Also in the sixth phase ESA's GD Dordain influenced the institutional change of ESA. He was the one conducting a scenario assessment and continued to push the debate on the agenda, while various Member States (e.g. Germany) started to consider the debate as "a dead horse" and "a cold potato" (Wörner, 2012).

4.2.4.3 Groups

During the first phase under discussion, Amaldi and Auger, who were members of an important epistemic community, and who were able to influence State interests, convincing governments of the need to work together, are a striking examples of group influence on institutional change or institutional foundation. However, the example of the foundation of ELDO also showed the contrary, i.e. the lack of group or experts' influence in the case of strong State interests. In fact, experts were sceptic and opposed to ELDO, fearing the development of civilian launchers would be conducted at the expense of their national research programmes. Also, engineers perceived Blue Streak as already obsolete and thus saw little point in taking it as a basis. They regarded the division of the management and development of the four stages between the different countries as doomed to fail. These doubts were however overcome by both lobbying activities of the aerospace industries, which in September 1961 established a supranational body called Eurospace with the aim to "promote the development of aerospace activities in Western Europe", and by political consideration (Krige, 1993a, p. 19).

During the second phase, no proofs of groups acting as change agents could be identified.

Similarly, the third phase is a prime example of the influence of epistemic communities on institutional change, as the Wise Men can be considered an example of epistemic communities triggering institutional change. Policy makers asked for expert advice given the highly technical nature of this policy field (Haas, 1992, p. 12). The advice of the Wise Men, as well as their call for "changes, especially in the relationship between ESA and the EU" (Bildt, Peyrelevade, & Späth, 2000, p. 9), is often considered a turning point in the history of space collaboration in Europe. While there were several calls for a comprehensive approach to space policy already prior to the Wise Men Report, it was only after this report that the EC started to work on a holistic approach instead of focussing on the rather sectoral benefit of space technology to community policies (Reuter, 2007, p. 12).

No evidence of groups acting as change agents could be found during the fourth, fifth and sixth phase.

4.2.4.4 Institutions/The European Integration Process

Institutions can act as change agents in two ways. First, they can undergo an undermining process leading to institutional change, which has been covered in the previous section on endogenous sources of institutional change. Secondly, one institution can trigger change in another one, something that the EU Commission is often said to do. Both of which can also be seen as "broad changes in the policy environment", which has been covered in the previous section on exogenous sources of institutional change.

As has been described previously, the European space system was subject to exogenous shocks. The question of the British membership for example played a profound role in the demise of ELDO during the first phase. So did the entrance of the EC/EU into the space area during the third phase. A continuous push from the Commission to gain more competences in the field can be observed since then. While it is often explained by relying on the spill-over thesis of Neofunctionalism, that space technology was critical to the achievement of several community policy objectives, the process of spill-over is "far from automatic" (Keohane & Hoffmann, 1991, pp. 18–9). Rather, it is driven by institutions. In fact, Neofunctionalism refers to the Commission as an example for the autonomous actions of supranational actors contributing to European policy making (Pierson, 1998, p. 29). There is near consensus in the literature on European governance that the preferences of the Commission are best described as highly pro-integrationist; seeking "more Europe" (Tallberg, 2003, p. 28). Majone (1996) suggests "what the Commission attempt to maximize is its influence, as measured by the scope of its competences" (p. 65). The more extensive the scope of the Commission's competences, the greater its influence in EU policy-making, and the better its chances of moving European integration forward (Ibid.). Moe (1990) refers to this as "autonomous action of European institutional actors" (p. 121 in Kassim & Menon, 2003, p. 130). This assumption can be proved by relying on the documents[15] published by the Commission during the third phase. The White Paper, a Commission document, is said to have placed "space in Europe's toolbox" (Commission of the European Communities, 2003, p. 8) out of economic interests. As a consequence, space assets became understood as tools necessary for the implementation of the EU policies (Suzuki, 2003, p. 185). Moreover, by establishing the strength and weaknesses[16] of Europe's space efforts, the White Paper justifies a "Community action" as "both possible and desirable" and refers to it as an "added value".[17] The provisions contained in policy documents at the time, all refer to a strategy of "synergy", "complementarity" and coordination" (Madders, 1997, p. 578). This is in line with the "concept of institutional complementarity" (Hall & Soskice, 2001) or "institutional coupling" (Madariaga, 2011, p. 25) as introduced by Sociological Institutionalism (cf. Chap. 2). Accordingly, institutions complement each other when the presence or absence of one institution affects the efficiency of the other (Hall & Soskice, 2001). Institutional change is the natural result of such a coupling, as a shift in one institution through this complementarity generates a shift in the other one (Madariaga, 2011, p. 25). The continuous reaction of ESA's DG Dordain can be seen as one example of such shifts. The case of ESA thus offers an example of the impact of the European integration process on existing institutional actors. Institutions like the Commission and ESA, once coexisting in harmony, become conflictual because new wants, new ideas, or new political interests produce instability in a previously complementary relationship (Norgaard, 1996, p. 47). The Commission has also enjoyed several windows of opportunities such as the dependence on the U.S. GPS system during the Balkan conflicts (Marta & Stephenson, 2016, p. 99). The fact that the Commission conducted an impact assessment shows

[15]For example European Parliament "Resolution on Community participation in space research" of 25 April 1979, OJ C 127 of 21 May 1979, p. 42 http://eur-lex.europa.eu/LexUriServ/LexUriServ.do?uri=OJ:C:1979:127:FULL:EN:PDF.

[16]"Europe's efforts remain in some respects selective: Europe holds a strong position in the areas it has targeted but it is less active and more dependent in others. Europe depends on the U.S. for some operations in space and it has not yet embarked on the development of very heavy or recoverable launch vehicles, a factor which may restrict its autonomy of exploitation of space in the future: Europe is still neither active enough nor sufficiently well organised when it comes to exploiting the applications of space technology." (Commission of the European Communities, 2003, p. 8).

[17]It is important to emphasise that the principle of subsidiarity enshrined in the Maastricht Treaty assumes that any intervention by the Community be justified by its added value (Jourdain, 1995).

its desperation to advance the debate about the roles of the EU and ESA in space policy. The shift of space as a domain from the DG Research and Development to the DG Enterprise and Industry, further highlights that Europe's engagement in space goes beyond research, having an important industrial dimension. Bearing in mind the Commission documents, which have been published during this phase, the Commission can be said to have exerted influence on institutional and policy outcomes. The dialectic of the Commission's documents also seems to show the Commission's independent preferences.

Particularly in the fourth phase the central role of the Commission in advancing the debate about defining the roles of the different players, especially EU and ESA in the space arena, and in identifying a governance model for the space domain could be identified. Through documents such as the Commission's Impact Assessment of the European Space Policy (Commission of the European Communities, 2007b), the Communication on the European Space Policy (Commission of the European Communities, 2007a) and the Commission's Space Policy Progress Report (Commission of the European Communities, 2008), the Commission increased the debate about the roles of EU and ESA. It is also possible to identify a change in the dialectic related to the EU-ESA relations. While in the previous phase, the Commission referred to a complementary rather undefined division of labour, it now defined its role as taking the lead in identifying user needs, while making ESA the technology development agency. For the first time, the Commission introduced the term "governance" in the debate about the competences of ESA and the Commission and space was also put into the context of global politics and developments, as a tool contributing to prosperity and global challenges. Moreover, this phase has been characterised among other things by the introduction of Galileo and Copernicus, the two flagship programmes of the EU's Commission. Both programmes can be seen as part of the "operationalisation" of the Commission as an actor in the space arena. The development of Galileo and Copernicus further highlighted questions related to governance and ESA-EU relations. At the same time, space applications became increasingly used in the area of security and environment. Given the two EU flagship programmes the European Commission also saw the need to found yet another space institution: The GNSS Supervisory Authority (GSA).[18] The GSA was restructured into the European GNSS Agency, ensuring the continuity of its activities. The change in the name gives further insights into the Commission's aspirations. So does the frequent reference by Commission personnel to the "space agency of the EU" for the GSA. In addition to the Commission the European Economic and Social Committee (EESC) and its EESC's Consultative Commission for Industrial Change (CCMI) on European Space Policy played a role during this phase, by proposing to renegotiate the FA as to achieve an "ever closer and more coordinated" (European Economic and

[18]Being established as a Community Agency on 12 July 2004, by Council Regulation (EC) 1321/2004, status amended in 2006 by Council Regulation (EC) No 1942/2006, the GSA is since Regulation (EU) No. 912/2010, which entered into force on 9 November 2010, called European GNSS Supervisory Authority but European GNSS Agency.

Social Committee, 2007, p. 5) relation between EU and ESA, given the European nature of space policy.

The Commission can also be considered a change agent in the fifth phase, with ESA being in a merely reactive position of institutional adjustment. It was the continuous push of the EU's Commission to introduce community competences for space in the founding treaties, which resulted in it being explicitly included in the Lisbon Treaty. In line with Norgaard (1996, p. 47), ESA and EU "once coexisting in harmony" have "become conflictual" due to instability caused by potentially doubled competencies. In the present case instability is caused by the shared competence in space policy as introduced by the Lisbon Treaty. The co-existence of EU and ESA in the space arena thus become unstable through the previous institutional complementarity. A shift in the Commission's competence resulted in a need for adaptation by ESA. In this context, it has to be highlighted that the Lisbon Treaty does neither explicitly give any competence to the EU's Commission, nor does it limit the Member States' sovereignty in developing their own national space policy. Instead, it is the Commission, which interprets the Lisbon Treaty as mandating it to develop a European Space Policy, thereby making "an absolute claim" on doing so (Wörner, 2012). The fifth phase is also the first during which the Commission explicitly outlines four possible scenarios for ESA's future, thereby also for the first time mentioning the option of bringing it within the framework of the EU following the example of EDA, as well as transforming it into an EU agency (European Commission, 2012, p. 4).

Similarly, it is possible to identify the Commission as prime change agent during the sixth phase. Not only did it task Roland Berger to conduct a cost-benefit assessment of the options, but when those were published, it concluded that "a conclusive choice among the options" (European Commission, 2012, p. 8) was not possible as the elements of analysis were "not (...) sufficient to justify" (Ibid.) such a choice. The main reason for this is the Commission's frustration about RB's recommendations.

4.2.5 Forms of Institutional Change in the Development of Space Policy in Europe

The following five different outcomes of institutional change have been identified in the literature under discussion:

- latent institutions become salient (Suzuki, 2003, p. 33) and are dissolved;
- institutions are put in the service of different ends (Ibid.);
- adjustment or learning: an existing institution is adjusted (Ibid.)
  - marginal adaptation to environmental changes without changing core principles;
  - gradual adaptation through partial renewal of institutional arrangements and limited redirection of core principles (Ebbinghaus, 2005, p. 17);
- bricolage: institutional change involving the recombination of old institutional elements

- substantive bricolage, involving the recombination of already existing institutional principles and practices, and
- symbolic bricolage, involving the recombination of symbolic principles and practises (Campbell, 2004, pp. 69–70);

• translation

- introduction of new institutional elements (Streeck & Thelen, 2005, p. 21)
- displacement: a new institution is founded (Ebbinghaus, 2005, p. 17; March & Olsen, 1989, p. 169; Suzuki, 2003, p. 33).

While theoretically it seems to be relatively easy to distinguish between the four different results of institutional change, i.e. institutional adjustment, bricolage, translation, displacement, the example of ESRO and ELDO in the first phase shows that the practice is relatively more complicated than these four options. While it is possible to consider the foundation of ESA based on ESRO's convention as institutional adjustment, the foundation of ESA has been conducted through a decision and not through mundane processes. At the same time, ESA was built on ESRO and ELDO, and thus involves old institutional elements such as the principle of geo-return of ELDO and many of the structural elements of ESRO. Thus, it could also be referred to as bricolage. However, ESA's convention also introduced new elements. Consequently, it would also be possible to refer to the process of change as translation with ELDO and ESRO ceased to exist.

ESA's introduction of the manned space programme as well as the Member States' manifestation to the independence of European space capabilities in the second phase can be considered as "institutional adjustment". The Member States' decision at the Rome Ministerial Council for the manifestation of autonomy and independence of European capabilities in space (ESA, 1985), which was driven by political and institutional reasons based on the experiences with the Aerosat programme with the U.S. can be said to exemplify a learning process as proposed by Olsen (1997, pp. 205–6). ESA benefitted from the experience gained in ESRO and during the ESC, in defining workable institutional procedures, which were incorporated in its Convention, producing a streamlined decision-making system (Madders, 1997, p. 174). This is what is referred to as "institutional heritage" or bricolage. Previous institutional forms constrain and influence subsequent choices of selecting alternative institutional forms (Florensa, 2004, p. 3). Florensa (2004, p. 9) refers to this phenomenon as "path dependence", in the sense that past institutions influence the direction of institutional change. During this phase, institutional change is incremental. Learning from the experience of ELDO and ESRO, ESA was founded with the objective to increase the efficiency of the European space efforts by making better use of the resources and to define a European space programme for exclusively peaceful purposes (cf. ESA, 1975). ESA's convention was further influenced by the progress of the U.S. and the USSR, leading to the inclusion of industrial policy as a purpose of ESA (Suzuki, 2003, p. 87).

Institutional change during the third phase took place through adjustment or learning, without any grand decision for change. ESA as an institution needed to adapt to the EU's new interest in the space field. Due to their previous co-existence as European players, both institutions had entered a relationship of complementarity, which became now threatened with the EU being interest in space policy. Thus, as a natural process ESA needed to adapt. In line with Olsen's (1997, pp. 209–10) understanding about institutions being transformed through "mundane processes of learning and incremental adaptation" and Thelen's (1999) "mechanisms of reproduction" (p. 387), ESA provides an example for the continuous adaption of institutions to changing circumstances. It also proves Olsen's (1997, p. 205–6) claim about efficient learning and adaptation processes circumventing the "accumulation of tensions and conflicts and radical change" (Ibid., pp. 209–10). The FA, which was negotiated by the JTF (Smith & Hörl, 2007) during the third phase can be considered as a compromise attempting to make the complementarity between the two institutions work and can thus be understood as institutional adjustment. In line with Madariaga (2011), institutional coupling or institutional complementarity is "a functional relationship between two institutions" with "a shift in one institution (…) gener[ating] through this complementarity a shift in the other" (p. 25).

In the following phases, ESA is adapting and adjusting to a changing environment, resulting from space being incorporated in the Lisbon Treaty as a shared competence.

4.3 Conclusions

Building on the existing theoretic accounts of institutional change, which were outlined in Chap. 2, this dissertation has proposed to structure the observations of institutional change by distinguishing five stages of institutional developments, consisting of

- agenda formation (Levy et al., 1995),
- institutional choice (i.e. selection and design) (Ibid.),
- operationalisation (Ibid.) or evolution (Florensa, 2004)
- redefinition (Ibid.) and
- crisis (Ibid.)[19]

and five facets of institutional change: (1) the degree of change (simple change or complex change), (2) the pace of change (incremental or evolutionary), (3) the sources of institutional change (endogenous, exogenous), activated by (4) change agents (States, individuals, groups or institutions) and (5) the outcome or form of institutional change (institutional adjustment, bricolage, translation, displacement).

In a second step, the historic development of space policy in Europe, has been outlined from the beginning of European space collaboration in 1959 until 2014.

[19]In line with this are two aspects to the evolution of an institution: first, the conditions under which institutions come about (agenda formation) and second, the process of competitive selection by relevant actors between different alternating institutional forms (institutional selection, choice or design) (cf. Florensa, 2004, pp. 9–10. cf. Hall & Taylor, 1996, p. 945).

A chronological approach has been adopted. Bearing in mind the importance of the time frame for the analysis of institutional change, six phases of institutional development of space policy in Europe have been identified, which in line with Aminzade (1992) and Campbell (2004, pp. 44–6) were distinguished by critical events.[20] This last chapter sought to combine the theoretical and empirical part of this dissertation. To do so, the six phases of the development of space policy in Europe were analysed in terms of the stages of institutional development and the five facets of institutional change. Particular attention was paid to the impact of the European integration process.

4.3.1 The Degree of Institutional Change of ESA

Given much academic and political debate on the role of ESA as part of the space governance structure in Europe, one might ask "To what extend has ESA changed institutionally so far?". By applying the identified facets of institutional change to the case of ESA, it becomes possible for the first time to provide a structured answer to this question. This dissertation has relied on Hall (1993, p. 278) and Campbell (2004, pp. 6, 48) to qualify the degree of change. Throughout the whole period of institutional development of ESA under discussion, no simple change in instruments and means, complex change or regulative change could be observed. Thus, one could say ESA has, if at all, changed in the cultural cognitive dimension, i.e. in its assumptions about reality and the frames through which it is perceived.

4.3.2 The Impact of the European Integration Process on the Development of Space Policy in Europe

The dissertation also sought to advance the debate about the impact of the European integration process on existing institutional actors, thereby making a theoretical contribution to our understanding of when supranational actors are able to advance their own agenda. The answer to this is twofold. First, the European integration process can be considered an exogenous sources for institutional change (cf. Sect. 4.2.3). Second, the European Commission as an institution can be considered a change agent (cf. Sect. 4.2.4.4). Indeed, in five out of six phases the impact of the European integration process was identified as an exogenous source for institutional change and in four out of six phases the Commission was found to have acted as a change agent. Merely in the second phase discussions in the context of the wider politics of European integration did not have an impact on the space field, which can be explained by Europe's integration ambitions coming to a halt in the 1960s and 1970s

[20]Accordingly, critical events may mark the appropriate time frames for analysis (e.g. electoral, budget, legislative or other cycles).

within the EC, when the Franco German differences caused a stalemate and put a stop to the efforts to achieve a common design for nuclear reactors.

4.3.3 Stages of Institutional Development as a Way to Structure the Observations on Institutional Change

One way to structure the observations of institutional change that was proposed by this dissertation was, to rely on Levy et al.'s (1995) and Florensa's (2004) stages of institutional development. The idea was to specify the time frame over which institutions are tracked, to determine how much institutional change has occurred. This is something, which has been called for by Campbell (2004, pp. 41–3) for example. Having applied these stages of institutional development to the case of ESA, the question arises, whether this proofed a fruitful endeavour. Indeed, Levy et al. (1995) and Florensa (2004) stages of institutional development were perceived as a good starting point and were considered as helpful in guiding the analysis of institutional change. They offer a way to structure the observations.

Moreover, given Sandholtz & Stone Sweet (1998, p. 16) claim about institutions being in constant evolution, one of the research questions of this dissertation was "Which stages of institutional development can be identified?". The case study under analysis has shown that the identified phases can be complemented by a possible sixth phase of institutional development. Following a period of redefinition is not necessarily a crisis but there can also be a phase of "consolidation". Alternatively, an institution can also be consolidating instead of experiencing a crisis. Moreover, the case study has shown that the phases of "consolidation" and "revision" are dominant. Founding phases and crisis are rather rare. This can also be seen in Fig. 4.1. One might conclude from this observation, that institutions are constantly revising and consolidating.

The case study has also shown that in practice, it can be difficult to determine when one of the phases starts and ends. They should thus be considered on a continuum.

Moreover, it was hoped to see whether the reasons for the selection of a particular institution, which have been identified by Pollack (1996, p. 434), Wendt (2001, pp. 1023–4) and Wallace (2000b, pp. 43–4) as rational expectations about costs and benefits (Wendt, 2001, p. 1043) and normative considerations of what is appropriate (Ibid., pp. 1023–4), can be identified in the case study or whether those institutions were chosen unintentionally as suggested by Héritier (2007, p. 1). Furthermore, it was hoped to find out, in how far the founders of the institutions were constrained by the availability of institutional alternatives and borrowed from prior institutions as suggested by Jupille (2004, p. 1) and Scott (2008, p. 94). Also it was hoped to discover in how far contextual factors, derived from the broad circumstances, functional factors, derived from some of the core functions of politics and demands to be met, and institutional factors, i.e. opportunities and constraints, influencing in how far institutions are able to channel and to structure the behaviour of political actors from

Phases under Conside-ration	Stages of institutional development					
	Agenda Formation	Institutional Choice	Operationali-sation or Evolution	Crisis	Consoli-dation	Redefinition
Phase 1 (1959-1974)	✓	✓	✓	✓	-----	✓
Phase 2 (1975-1986)	-----	✓	✓	-----	-----	✓
Phase 3 (1986-2004)	-----	-----	-----	-----	-----	✓
Phase 4 (2004-2009)	-----	-----	-----	-----	✓	-----
Phase 5 (2009-2012)	-----	-----	-----	-----	-----	✓
Phase 6 (post 2012)	-----	-----	-----	-----	-----	✓

Fig. 4.1 Phases of the development of space policy in Europe analysed in terms of stages of institutional development

the participating countries (Wallace, 2000b, pp. 43–4) were relevant in the process of institutional selection in the case of ESRO, ELDO and ESA. Indeed, ELDO, ESRO and ESA have been selected intentionally on the basis of rational expectations about the costs and benefits expected, through a process of intergovernmental bargaining on the basis of national preferences and other motivational factors. None of them came about unintentionally. Existing institutions, such as CERN, EURATOM and NATO, inspired the institutional choice of the founders of ESRO and ELDO. ESA benefitted from the experience of ESRO and ELDO and the pitfalls in their institutional design. Thus, considering Jupille (2004, p. 1) and Scott (2008, p. 94), it can be said to be right, as ESRO, ELDO and ESA borrowed from prior existing institutions. ELDO was created on the basis of political will, while ESRO was created given the growing recognition among the scientific community, of the stakes involved in space (Gaubert & Lebeau, 2009, p. 37). These findings are further visualised in Fig. 4.2.

It was also hoped to see whether the conditions for the foundation of an institution, i.e. the "problem condition", the "cognitive condition", the "hegemonic condition" (Rittberger & Zangl, 2006, p. 25), "windows of opportunity", and breaking points, which have been identified based on Rittberger and Zangl (2006, p. 25), DiMaggio (1988, p. 14) and Olsen (1997, pp. 209–10) were present when ESRO, ELDO and ESA were founded. Indeed, the proposed conditions necessary for the foundation of an institution could be identified in the case of ESRO and ELDO. It was not possible to identify the hegemonic condition in the case of ESA. These findings are further visualised in Fig. 4.3.

Phases under Conside-ration	Institutional Choice			
	Rational Expectations	Normative considerations	Unintentional choice	Institutional alternatives
Phase 1 (1959-1974)	✓	✓	✓	✓
Phase 2 (1975-1986)	✓	✓	-----	✓
Phase 3 (1986-2004)	-----	-----	-----	-----
Phase 4 (2004-2009)	-----	-----	-----	-----
Phase 5 (2009-2012)	-----	-----	-----	-----
Phase 6 (Post-2012)	-----	-----	-----	-----

Fig. 4.2 Phases of the development of space policy in Europe analysed in terms of reasons for the selection of a particular institution

Phases under Conside-ration	Conditions for the Foundation of an Institution				
	Problem condition	Cognitive condition	Hegemonic condition	Window of opportunity	Breaking point
Phase 1 (1959-1974)	✓	✓	✓	✓	✓
Phase 2 (1975-1986)	✓	✓	-----	-----	✓
Phase 3 (1986-2004)	-----	-----	-----	-----	-----
Phase 4 (2004-2009)	-----	-----	-----	-----	-----
Phase 5 (2009-2012)	-----	-----	-----	-----	-----
Phase 6 (Post-2012)	-----	-----	-----	-----	-----

Fig. 4.3 Phases of the development of space policy in Europe analysed in terms of conditions for the foundation of an institution

4.3.4 Facets of Institutional Change as a Way to Structure the Observations on Institutional Change

Some of the research questions posed at the beginning of this dissertation were "What are the sources for institutional change?", "Which factors are most important in describing institutional change?", "Which outcomes of institutional change are possible?" and "How can different cases of institutional change best be compared?".

To answer these, this dissertation has identified five "facets of institutional change" based on the existing literature on institutional change. A facet was understood as "one of several parts",[21] being synonymous to "aspects".

To make a theoretical contribution to our understanding of institutional change and to structure the debate on this topic, the question arises whether the proposed facets of institutional change proofed to be suited, to structure the debate on institutional change. Some of which might still be too superficial to qualify institutional change.

4.3.4.1 Degree of Institutional Change

First, the degree of change as a dimension to qualify institutional change proofed rather difficult to distinguish. The simple differentiation between simple change, i.e. in instruments and means, and complex change, i.e. in goals and ends (cf. Hall, 1993, p. 278) proved to be too superficial, as all phases under discussion qualified for simple change. Even when a new institution was newly founded, it had the same goals and ends, and thus no complex change could be identified. At the same time, the more differentiated distinction in changes in the regulative, normative and cultural cognitive institutional dimension proved insufficient to grasp all facets of the case under discussion. In practice, it seemed to be rather difficult to identify change in the cultural cognitive dimension, which does not seem to be an objective characteristic, but can be rather subjective in nature. Further research has to be conducted as to find a suitable theoretic framework to differentiate the degree of change further.

4.3.4.2 Pace of Institutional Change

Secondly, the pace of institutional change, which was distinguished in slow, incremental through small adjustments, gradual or evolutionary in a continuous manner and rapid revolutionary change through large steps in a discontinuous manner, resulting in a "steep transformation", breakdown and replacement of old institutional arrangements (Campbell, 2004, p. 33; Madariaga, 2011, pp. 17, 29; Roland, 2004, pp. 116–7), proved to be a sufficient theoretic framework to qualify institutional change. Incremental changes as suggested by Historical Institutionalists (i.e. Campbell 2004, p. 48; March and Olsen 1989, p. 169; North 1990, p. 73; Pollack 1996, p. 441), seem to provide for the "main story", as four out of six phases under analysis can be considered to have changed in an incremental manner. Only the founding of ESA (phase 1 and phase 2), replacing ELDO and ESRO, can be considered a large step in a discontinuous manner. The case of space policy also highlighted how small incremental changes can culminate in evolutionary change, an example being the incorporation of space policy in the Lisbon Treaty as a shared competence.

[21] (cf. Longman Dictionary of Contemporary English).

4.3.4.3 Sources of Institutional Change

Third, it was proposed to distinguish between endogenous sources of institutional change (i.e. Greif and Laitin 1994; Streeck and Thelen 2005; Young 2010), such as self-undermining processes (Greif & Laitin, 1994, p. 634), institutional design (Héritier, 2007; Streeck & Thelen, 2005, p. 19) or internal contradictions (Young, 2010, p. 174), ideological innovation, learning processes through the acquisition of knowledge and skills and changes in tastes or preferences of the actors involved (Florensa, 2004, p. 8), and exogenous (i.e. Krasner, 1988, p. 67; Pollack, 1996, p. 441; Scott, 2008; Steinmo et al., 1992; Suzuki, 2003) sources of institutional change. Those would be "critical junctures or punctuation" (cf. Historical Institutionalism), i.e. changes in the attribution of natural resources, extreme climatic events or natural disasters,[22] changing preferences of supportive/influencing governments (Pollack, 1996, p. 441), which might result from unfulfilled expectations, shifts in power distributions of previously supportive coalitions or the appearance of new actors (Ibid., p. 438; Florensa, 2004, p. 18; Scott, 2008, p. 198), changing voter preferences, leading to new political alignments, lobby of business interest (Scott, 2008, p. 196), broad changes in the policy environment of social, political or economic nature (Florensa, 2004, p. 18; Pollack, 1996, p. 438),[23] and improvements in the quality of information about either the operation of the institution or the environment within which it operates (Pollack, 1996, p. 438). It was expected to find both endo- and exogenous sources, when analysing the case study. In four out of six phases under analysis, both endogenous and exogenous sources of institutional change could be identified. Since the second phase mainly deals with the "operationalisation" and founding of ESA naturally neither endogenous nor exogenous sources for change could be identified. The differentiation between endogenous and exogenous sources of change is difficult during the fourth phase as this is characterised by the resolutions issued by the Space Council consisting of both the ESA Council and the EU Competitiveness Council. The two main exogenous sources, which were identified in the six phases under analysis, were the impact from the European integration process and from the geopolitical context mainly coming from the U.S. In fact, in five out of six phases under discussion, institutional change can be said to have resulted from the geopolitical context. This will be summarised in Fig. 4.4.

While such a distinction proved useful for guiding the analysis and making different cases of institutional change comparable, the question arises whether the mere distinction in exogenous and endogenous sources is fruitful, if both seem to be present in most phases under discussion. Also, there is a hen-egg-problem inherent in the distinction, since in practice it can be hard to distinguish the original source, be it the self-undermining process (endogenous source) or the broad changes in the pol-

[22] For more information on "critical junctures" refer to the following literature (cf. Campbell, 2004, p. 26; Christiansen & Vanhoonacker, 2008; Florensa, 2004, p. 18; Hall & Taylor, 1996, p. 942; Krasner, 1984, p. 235 in Ikenberry, 1994, p. 9; Horak, 2007, p. 21; March & Olsen, 2006, p. 12).

[23] For example, changes in the markets, technological innovations or legislative changes (Steinmo et al., 1992, p. 16).

Phases under Consideration	Endogenous Sources	Exogenous Sources
Phase 1: The Historic Evolution of ESA (1959-1974)	• enshrined in institutional design (→characteristics of founding community of ESRO) • inadequate financial resources (ESRO & ELDO) • malfunctioning established structure (agenda crowded by trivialities → ESRO) • self-undermining process → crisis of confidence of ESRO • difficulties compounded by the principle of juste retour (ELDO) • lack of an institutional mechanism for political discussion (ELDO) • lack of an institutional mechanism for coordinating activities btw. ELDO & ESRO	• broad changes in the policy environment particularly resulting from the European integration process • superpower rivalry • changing preferences of supportive governments • critical junctures • unfulfilled expectations
Phase 2: From the founding of ESA to the entering of the EC in the space domain (1975-1986)	• none	• U.S. / geopolitical context (autonomy / independence)
Phase 3: From the first steps of the EC to the ESA-EU Framework Agreement (1986-2004)	• revision process (WG, Wise Men, Agenda 2007): questioning mission & evaluating its role vis-à-vis other institutions	• COM as competitor interested in space field due to applications related to community goals • U.S. / GPS / limited civilian use during Kosovo crisis; autonomous European navigation system needed
Phase 4: From the Framework Agreement to the Lisbon Treaty (2004-2009)	• contd. revision process and institutional adjustment (i.e. Agenda 2011; ESA's DG Proposal for the ESP)	• window of opportunity: review of EU legal framework • European industry to counteract U.S. corporations →need to work together • U.S: and China strong in R&D →Europe needs to work together • New U.S. space policy following elections of Obama
Phase 5: Post-Lisbon Treaty developments (2009-2012)	• contd. revision process in the light of the Lisbon Treaty (i.e. political declaration "Towards the ESA that best serves Europe")	• window of opportunity: review of EU legal framework • European industry to counteract U.S. corporations →need to work together • U.S: and China strong in R&D →Europe needs to work together • new U.S. space policy following elections of Obama
Phase 6: Post-ESA-EU Declaration of November 2012 (Post-2012)	• contd. revision process (i.e. scenario assessment commissioned by ESA's DG)	• contd. push from COM

Fig. 4.4 Phases of the development of space policy in Europe analysed in terms of endo- and exogenous sources of institutional change

icy environment (exogenous source), which might have led to the self-undermining process in the first place. Conditions under which endogenous or exogenous sources preside could not be identified. Additionally, academics did not clearly define whether they considered changing preference of Member States of an intergovernmental institution as exogenous factors or endogenous sources of institutional change. This dissertation decided to consider them as exogenous sources.

	Phase 1 (1959-1974)	Phase 2 (1975-1986)	Phase 3 (1986-2004)	Phase 4 (2004-2009)	Phase 5 (2009-2012)	Phase 6 (Post-2012)
States	Founding: UK, FRA, GER; Crisis: Spain, Italy	ESA MS	UK Presidency	MS of ESA and EU through Space Council	FRA + GER	MS of ESA and EU
Individuals	Auger, Amaldi, Massey, Thornycroft, Blamont, Causse, Barnier, Spacey, Puppi, Bondi, Aubinière	none	Rodota, Lord Sainsbury, Delors, Contzen, Allgeier	Dordain	Dordain, Wörner	Dordain
Groups	Scientific Community	none	Wise Men	none	none	none
Institutions	European Integration Process	none	COM	COM	COM	COM

Fig. 4.5 Phases of the development of space policy in Europe analysed in terms of change agents

4.3.4.4 Change Agents

Fourth, it was suggested that sources of institutional change "have to be activated by agents willing to generate change" (Madariaga, 2011, p. 24), such as States (and their ministries, representatives), individuals, groups (interest groups and communities of experts) or institutions (administrative staff of international organisations; parliamentary assemblies of international organisation) (Rittberger & Zangl, 2006, p. 78). The case under discussion showed an existence of all four groups of change agents across the six phases under analysis, i.e. States (in all six phases), individuals (five out of six phases), groups such as epistemic communities (two out of six phases) and other institutions such as the Commission (in four out of six phases) could be identified to have acted as a change agent.

A summary of the different change agents in the different phase of institutional development can be found in Fig. 4.5.

States

In line with Doleys (2000) one of the research questions was "in what situations governments can be expected to dominate policy-making, and under what conditions supranational actors can be expected to affect change independently of and even in the face of opposition from governments?" It was expected to find out that States acted as change agents activating institutional change in all six phases under discussion, given that space policy is a matter of national sovereignty, a symbol of scientific and technical capability, an index of political power and a component of military strategy, and thus governed by States. Indeed, States were found to have acted as change agents in all six phases under discussion.

Moreover, the current status quo of the inter-institutional relationship between ESA and the EU is a result of intergovernmental bargaining at the ESA Ministerial Council in November 2014. Keeping everything as it is, could be regarded as the "lowest common denominator". Several authors refer to institutions chosen by Member States as "equilibrium institution", meaning that, there is "no possible coalition of actors with an institutional right to change the existing institutions, has an incentive to do so" (Pollack, 1996, p. 437). This is what Scharpf came to refer to as the "Joint-Decision Trap". Accordingly, a given institutions, once instituted, tends to remain in place, rigid and inflexible, even in the face of a changing policy environments due to intergovernmentalism and decisions by unanimity (Pollack, 1996, p. 440). Moreover, it is clear, that Europe can only compete with the U.S., Russia and China when working together. However, given that space is a matter of national sovereignty, States do not want to give up their influence in European space collaboration by integrating ESA as a EU agency into the EU. This would result in ESA being subordinated to a DG of the Commission and thus space collaboration would be decided by a non-elected supranational institution. Instead, ESA is currently being governed by the ESA Ministerial Council, composed of the national ministers in charge of space activities of the ESA Member States. Thus, ESA by its very nature is an intergovernmental organisation (Wörner, 2013).

The case of the foundation of ESRO and ELDO further highlights when States influence presides over the one of individuals or experts. While ESRO can be said to be the result of the call of the scientific community and mainly Amaldi's and Auger's power of negotiations, ELDO was the result of intergovernmental negotiations and not scientific considerations. This can be explained by the fact that ELDO dealt with launchers, which can be considered a question of national sovereignty. Thus, one could say that States' influence presides over the influence of individuals (experts) in questions of national sovereignty. In order to generalise this conclusion, it needs to be tested in different case studies.

Individuals

The case study shows that intergovernmental bargains, as proclaimed by Neo-Realism, were not the only driving force of European space collaboration and institutional change. Scientists, engineers, and other experts were also involved in the creation and change of the institutions of European space collaboration. Both Neorealism and Neoliberal institutionalism, which neither account for the role of epistemic communities nor for the role of individuals, like Auger and Amaldi, are able to fully explain the case of European space collaboration. On the contrary, the case shows that individuals can play an important if not essential role in the agenda formation by highlighting the need for cooperation. Fligstein (1997a, 1997b, 2010), North (1990, p. 73) and Sabatier and Jenkins-Smith (1996) (Sabatier, 1998, p. 109)[24] have considered so-called "institutional entrepreneurs" as the main source for institutional change, defining them as individuals, who have their own preferences, creating opportunities for innovation and institutional change (Thornton, Jones, & Kury, 2005, p. 129). Their assumptions can be underlined by the case at hand.

In line with Fligstein (1997a, p. 398), one area of particular interest of this dissertation was to analyse, whether in times of uncertainty or when highly technical issues are being discussed, skilful individuals with high social capital and the ability to access decision makers are able to advance their own agenda and act as change agents or can take the lead in founding institutions. Since space policy is of highly technical nature, it was expected that individuals were able to act as change agents in all phases under discussion. Individuals were found to have acted as change agents in five out of six phases under discussion. Only during the foundation process of ESA, no influence of individuals could be identified. Auger and Amaldi were considered as credible in the policy field, given their experience in founding CERN. Their social capital as well as their success in framing the story about the need for European space collaboration can be attributed to their past in CERN. In line with this, Campbell (2004) has highlighted the role of individuals as change agents in the context of bricolage; i.e. institutional change involving the recombination of old institutional elements (Campbell, 2004, p. 28; Lanzara, 1998, p. 26). In line with Campbell (2004), it is not surprising that Auger and Amaldi, given their experience with CERN, recombined institutional elements in innovative ways, leading to the institution building and the foundation of ESRO.

Groups

Another area of particular interest of this dissertation was, whether in times of uncertainty or when highly technical issues are being discussed, policy-makers' would demand information, which are provided by expert groups such as epistemic communities (Haas, 1992, p. 12). In two out of six phases, individuals who were part of an epistemic community acted as change agents and are thus considered as groups acting as change agents. As has been shown, the scientific community acted as one of the driving sources for the foundation of ESRO and during the third phase, when the Commission tasked the Wise Men to provide independent advice for promoting the evolution of the ESA-EU relationship and the necessary institutional change. Both are examples of the influence of groups in the context of institutional change.

---

[24] cf. Dudley and Richardson (1999, p. 227).

Institutions and Institutional Interrelations

According to theoretic accounts such as Neofunctionalism, Supranational Governance and Rational Choice Institutionalism supranational institutions exert significant independent influence on institutional and policy outcomes (Hix, 2005, p. 16). They act as driving forces (Pierson, 1998, p. 29), play an autonomous role (Andersen, Eliassen, & Sitter, 2001, p. 28) and can act as change agents.

Since the influence of supranational actors is not a new phenomenon, the interesting question is, when are supranational actors able to advance their own agenda, which was thus one of the research questions of this dissertation. It was expected that the Commission would be able to advance its own agenda regarding topics were it has a great authority, legitimacy, or "institutional standing" and "institutional embeddedness", when it has the control of information and knowledge, when it is able to abuse information asymmetries, to frame or shape a proposal in such a way, that it is securing approval even in the light of Member State(s)'s opposition, or when it acts together with powerful allies such as States. None of which could be empirically proven.

Nevertheless, in four out of six phases[25] the Commission, an institution, can be identified to have acted as a change agent. This is striking, as the Commission was a new actor to the field of space policy and thus lacked authority, legitimacy, or institutional standing in this policy field. States were also not particularly fragmented, when the Commission entered the policy field. Moreover, being of interest to national sovereignty, the space field was expected to be prone to States' influence and not influence of a supranational actor. The Commission's influence can partly be explained by enjoying several windows of opportunities such as the Kyoto Protocol and the Balkan conflicts (cf. Sects. 3.3.3 and 4.2.3.2) (Marta & Stephenson, 2016, p. 99), which the Commission was able to use to frame space policy in a way, that it is a tool to reach the Community's objectives. This way it justified its extended role in this policy field. Another explanation could be the Commission's right of initiative enshrined in the Community legislation.

4.3.4.5 Forms of Institutional Change

Fifth, while theoretically it seems to be relatively easy to distinguish between the five different outcomes of institutional change, as suggested by the literature under discussion, i.e. salience/dissolution, different services/ends, institutional adjust-

[25] In the first phase it was the European integration process as a contextual factor influencing the field of space policy and the development of ESA rather than the institution Commission. In the second phase, the founding phase of ESA, no influence of the Commission could be identified, which can be explained by the Franco German differences to the common design for nuclear reactors, which caused a stalemate and made Europe's integration ambitions coming to a halt in the 1960s and 1970s. Since the fourth phase under analysis, a central role of the Commission in advancing the debate about defining the roles of the different players, especially EU and ESA in the space arena, and in identifying a governance model for the space domain could be identified.

ment/learning, bricolage and translation, the case study showed, that in practice, it is relatively more complicated to categorise institutional change according to these options. While the first and the second phase seem to cover all possible outcomes, apart from the institution being put in the service of different ends, the other phases can all be characterised as institutional learning and adaptation.

References

Aminzade, R. (1992). Historical sociology and time. *Sociological Methods & Research, 20*, 456–480.

Andersen, S. S., Eliassen, K. A., & Sitter, N. (2001). Formal processes: EU institutions and actors. In S. S. Andersen & K. A. Eliassen (Eds.), *Making policy in Europe* (pp. 20–43). London: SAGE Publisher.

Battrick, B. (Ed.). (2003). *Agenda 2007—A document by the ESA director general* (BR-213). Retrieved from: https://www.espi.or.at/images/documents/agenda%202007%20-%20a%20document%20by%20the%20esa%20director%20general%20.pdf. Accessed March 19, 2017.

Bildt, C., Peyrelevade, J., & Späth, L. (2000). *Towards a space agency for the European Union* [Wise-Men-Report]. Retrieved from: http://www.espi.or.at/images/documents/the%20wise%20men%20report.pdf. Accessed January 18, 2012.

Borras, S. (2006). *The innovation policy of the European Union: From government to governance*. Cheltenham: Edward Elgar.

Campbell, J. (2004). *Institutional change and globalization*. Princeton & Oxford: Princeton University Press.

Cheli, S., & Schrogl, K. (1999). Reshaping European space activities. *Space Policy, 15*(2), 61–66.

Christiansen, T., & Vanhoonacker, S. (2008). At a critical juncture? Change and continuity in the institutional development of the Council Secretariat. *West European Politics, 31*(4), 751–770.

Commission of the European Communities. (2003). *Space: A new European frontier for an expanding union—An action plan for implementing the European space policy* (COM (2003) 673) [White Paper]. Brussels: European Communities.

Commission of the European Communities. (2008). *European space policy progress report* (COM (2008) 561 final). Brussels: European Communities.

Commission of the European Communities. (2007a). *European space policy* (COM (2007) 212 final). Retrieved from http://eur-lex.europa.eu/legal-content/EN/TXT/?uri=CELEX:52007DC0212. Accessed March 13, 2017.

Commission of the European Communities. (2007b). *Impact assessment of the European space policy* (SEC (2007) 505 draft). Brussels: European Communities.

Council of the European Union. (2009). *Resolution on the contribution of space to innovation and competitiveness in the context of the European economic recovery plan, and further steps* (10500/09). Brussels: European Communities.

Creola, P. (2001). Some comments on the ESA/EU space strategy. *Space Policy, 17*, 87–90.

Dehousse, R., & Magnette, P. (2006). Institutional change in the EU. In J. Peterson & M. Shackleton (Eds.), *The institutions of the European Union* (pp. 17–34). Oxford: OUP.

DiMaggio, P. (1988). Interest and agency in institutional theory. In L. G. Zucker (Ed.), *Institutional patterns and organizations: Culture and environment* (pp. 3–23). Cambridge: Ballinger Publishing.

Doleys, T. (2000). Member States and the European Commission: Theoretical Insights from the new Economics of Organization. *Journal of European Public Policy, 7*(4), 532–553.

Dudley, G., & Richardson, J. (1999). Competing advocacy coalitions and the process of "frame reflection": A longitudinal analysis of EU steel policy. *Journal of European Public Policy, 6*(2), 225–248.

Dupas, A., Janischewski, S., von Kries, W., & Schrogl, K. (2001). A Franco-German view of Europe's ambition in space for the 21st century. *Space Policy, 17,* 103–110.
Ebbinghaus, B. (2005). *Can path dependence explain institutional change—Two approaches applied to welfare state reform*. MPIfG Discussion Paper 05/2. Retrieved from: http://www.mpifg.de/pu/mpifg_dp/dp05-2.pdf. Accessed March 17, 2017.
European Commission. (2012). *Establishing appropriate relations between the EU and the European Space Agency* (COM (2012) 671 final). Brussels: European Communities.
European Economic and Social Committee. (2007). *Opinion of the European Economic and Social Committee on the COM* (COM (2007) 212 final). OJ C 126 of 25 June 2008. Brussels: European Communities.
ESA. (1975). *Convention for the establishment of a European Space Agency*. Retrieved from: http://www.esa.int/About_Us/Law_at_ESA/ESA_Convention. Accessed March 19, 2017.
ESA. (1985). *Council resolution on the long-term European space plan* (ESA/C/LXVII/res1 (final)). Retrieved from: http://download.esa.int/docs/LEX-L/Ministerial-Council-Resolutions/Meeting-C-M067(Rome,30-31Jan1985)/ESAC-MLXVIIRes.1(Final).pdf. Accessed March 19, 2017.
ESA. (1999). Resolution shaping the future of Europe in space (ESA/C-M/CXLI/Res.1 (final). *ESA Bulletin, 98*, 20–25.
ESA. (2007). *Resolution on the European space policy—ESA Director General's proposal for the European space policy* (ESA BR-269 of 2007). Retrieved from: http://esamultimedia.esa.int/multimedia/publications/BR-269/. Accessed March 19, 2017.
ESA. (2012). *Political declaration towards the European Space Agency that best serves Europe* (ESA/C-M/CCXXXIV/Res.4 (final)). Naples: ESA.
Eurospace. (1995). Space: A challenge for Europe. *Space Policy, 11*(4), 227–232.
Fligstein, N. (1997a). Social skill and institutional theory. *American Behavioral Scientist, 40*(4), 397–405.
Fligstein, N. (1997b). *Social skill and the theory of fields*. Retrieved from: http://www.irle.berkeley.edu/culture/papers/Fligstein01_01.pdf. Accessed March 13, 2017.
Fligstein, N. (2010). Institutional entrepreneurs and cultural frames—The case of the European Union's single market program. *European Societies, 3*(3), 261–287.
Florensa, M. C. (2004). *Institutional stability and change. A logic sequence for studying institutional dynamics*. Paper presented at the Tenth Biennial Conference of the International Association for the Study of Common Property (IASCP). Oaxaca, Mexico. Retrieved from: https://dlc.dlib.indiana.edu/dlc/bitstream/handle/10535/1823/CostejaFlorensa_Institutional_040525_Paper220.pdf?sequence=1. Accessed March 13, 2017.
Gaubert, A., & Lebeau, A. (2009). Reforming European space governance. *Space Policy, 25,* 37–44.
Greif, A., & Laitin, D. D. (1994). A theory of endogenous institutional change. *American Political Science Review, 98*(4), 633–652.
Haas, P. M. (1992). Introduction: Epistemic communities and international policy coordination. *International Organization, 46*(1), 1–35.
Hall, P. A. (1993). Policy paradigms, social learning, and the state: The case of economic policy-making in Britain. *Comparative Politics, 25*(3), 275–296.
Hall, P. A., & Soskice, D. (2001). *Varieties of capitalism: The institutional foundations of comparative advantage*. Oxford: OUP.
Hall, P. A., & Taylor, R. C. R. (1996). Political science and the three new institutionalisms. *Political Studies, XLIV*, 936–957.
Héritier, A. (2007). *Explaining institutional change in Europe*. New York: OUP.
Hesse, M., & Hornung, M. (2012). Neue Impulse der Europäischen Kommission für eine europäische Weltraumstrategie. *Integration, 35*(3), 183–199.
Hix, S. (2005). *The political system of the European Union*. Houndmills: Palgrave.
Hobe, S., Kunzmann, K., & Reuter, T. (2006). *Forschungsbericht ESA – EU: Rechtliche Rahmenbedingungen einer zukünftigen kohärenten Struktur der europäischen Raumfahrt*. Kölner Schriften zum Internationalen und Europäischen Recht Bd. 13. Berlin: LIT Verlag.

Horak, M. (2007). *Governing the post-communist city: Institutions and democratic development in Prague*. Toronto: University of Toronto.

Hörber, T. (2016a). The European Space Agency and the European Union. In T. Hörber & P. Stephenson (Eds.), *European space policy* (pp. 53–65). London & New York: Routledge.

Hörber, T. (2016b). Chaos or consolidation? Post-war space policy in Europe. In T. Hörber & P. Stephenson (Eds.), *European space policy* (pp. 15–29). London & New York: Routledge.

Ikenberry, G. J. (1994). *History's heavy hand: Institutions and the politics of the state*. Paper prepared for a conference on New Perspectives on Institutions, University of Maryland. Retrieved from: https://scholar.princeton.edu/sites/default/files/HistorysHeavyHand_0.pdf. Accessed March 13, 2017.

Jourdain, L. (1995). Space in Europe: New stakes, new structures? *Space Policy, 11*(2), 87–88.

Jupille, J. (2004). *Procedural politics—Issues, influence, and institutional choice in the European Union*. Cambridge: Cambridge University Press.

Kassim, H., & Menon, A. (2003). The principal-agent approach and the study of the European Union: Promise unfulfilled. *Journal of European Public Policy, 10*(1), 121–139.

Keohane, R., & Hoffmann, S. (1991). *The new European community—Decision-making and institutional change*. Boulder/San Francisco/Oxford: Westview Press.

Keohane, R., & Nye, J. (2001). *Power and interdependence*. New York: Longman.

Köpping Anthanasopoulos, H. (2016). Europe's wilderness—The council's frames on space policy. In T. Hörber & P. Stephenson (Eds.), *European space policy* (pp. 82–97). London & New York: Routledge.

Koremenos, B., Lipson, C., & Snidal, D. (2001). The rational design of international institutions. *International Organization, 55*(4), 761–799.

Kraft, G. (2012). *For one German opinion*. DLR Newsletter Countdown. Special Edition of December 2012. Bonn: DLR.

Krasner, S. (1988). Sovereignty—An institutional perspective. *Comparative Political Studies, 21*(1), 66–94.

Krige, J. (1992). *The prehistory of ESRO 1959/60 From the first initiatives to the formation of the COPERS*. ESA History Study Report HSR-1. Nordwijk: ESA.

Krige, J. (1993a). *Europe into space: The auger years (1959–1967)*. ESA History Study Report HSR-3. Nordwijk: ESA.

Krige, J. (1993b). *The early activities of the COPERS and the drafting of the ESRO convention (1961/62)*. ESA History Study Report HSR-4. Nordwijk: ESA.

Krige, J. (1993c). *The launch of ELDO*. ESA HSR-7. Nordwijk: ESA.

Krige, J. (1994). The European space system. In J. Krige & A. Russo (Eds.), *Reflections on Europe in space* (pp. 1–13). ESA HSR-11. Nordwijk: ESA.

Krige, J. & Russo, A. (2000). *A history of the European Space Agency: 1958-1987* (Vol. I). SP - 1235 of April 2000. Nordwijk: ESA.

Lanzara, G. F. (1998). Self-destructive processes in institution building and some modest countervailing mechanisms. *European Journal of Political Research, 33*, 1–39.

Levy, M. A., Young, O. R., & Zürn, M. (1995). The study of international regimes. *European Journal of International Relations, 1*, 267–330.

Madariaga, A. (2011). *Patterns of institutional change and external competitiveness in neoliberal and dependent political economies. The cases of Chile and Estonia*. Budapest: Central European University.

Madders, K. (1997). *A new force at a new frontier—Europe's development in the space field in the light of its main actors, policies, law and activities from its beginning up to the present*. Cambridge: Cambridge University Press.

Majone, G. (1996). *Regulating Europe*. London: Routledge.

March, J. G., & Olsen, J. P. (1989). *Rediscovering institutions: The organizational basis of politics*. New York: The Free Press.

March, J. G. & Olsen, J. P. (2006). Elaborating the 'New Institutionalism'. In R. A. W. Rhodes, S. A. Binder, & B. A. Rockman (Eds.), *The Oxford handbook of political institutions* (pp. 3–20). Oxford: OUP.
Marta, L., & Stephenson, P. (2016). Role of the European Commission in framing European space policy. In T. Hörber & P. Stephenson (Eds.), *European Space Policy* (pp. 98–113). London & New York: Routledge.
Moe, T. M. (1990). The politics of structural choice: Toward a theory of public bureaucracy. In O. E. Williamson (Ed.), *Organization theory: From Chester Barnard to the present and beyond* (pp. 116–153). Oxford: OUP.
Moravcsik, A. (1993). Preferences and power in the European community: A liberal intergovernmentalist approach. *Journal of Common Market Studies, 31*(4), 473–522.
Norgaard, A. (1996). Rediscovering reasonable rationality in institutional analysis. *European Journal of Political Research, 29,* 31–57.
North, D. C. (1990). *Institutions, institutional change and economic performance*. Cambridge: Cambridge University Press.
Olsen, J. P. (1997). Institutional design in democratic context. *The Journal of Political Philosophy, 5*(3), 203–229.
Peter, N. (2007). The EU's emergent space diplomacy. *Space Policy, 23,* 97–107.
Pierson, P. (1998). The path to European integration: A historical institutionalist perspective. In W. Sandholtz & A. Stone Sweet (Eds.), *European integration and supranational governance* (pp. 27–58). Oxford: OUP.
Pollack, M. A. (1996). The new institutionalism and EC governance: The promise and limits of institutional analysis. *Governance: An International Journal of Policy, Administration, and Institutions, 9*(4), 429–458.
Rathgeber, W. & Remuss, N. (2009). *Space security—A formative role and principled identity for Europe*. ESPI Report 16. Wien: ESPI.
Remuss, N. (2010). Space and security—Challenges for Europe. *Sicherheit und Frieden, 3,* 151–160.
Remuss, N. (2011). Space and security. In C. Brünner & A. Soucek (Eds.), *Outer space in society, politics and law* (pp. 519–554). Vienna: Springer.
Reuter, T. (2007). *Die ESA als Raumfahrtagentur der Europäischen Union - Rechtliche Rahmenbedingungen für eine institutionelle Neuausrichtung der europäischen Raumfahrt*. Köln: Carl Heymanns Verlag.
Rittberger, V., & Zangl, B. (2006). *International organization—Polity, politics and policies*. New York: Palgrave Macmillan.
Roland, G. (2004). Understanding institutional change: Fast-moving and slow-moving institutions. *Studies in Comparative International Development, 38*(4), 109–131.
Rosamond, B. (2000). *Theories of European integration*. Houndmills: Palgrave.
Sabatier, P. A. (1998). The advocacy coalition framework: Revisions and relevance for Europe. *Journal of European Public Policy, 5*(1), 98–130.
Sabatier, P. A., & Jenkins-Smith, H. S. (1996). *The advocacy coalition framework: An assessment*. Paper presented to the Annual Meeting of the American Political Science Association.
Sandholtz, W., & Stone Sweet, A. (1998). *European Integration and Supranational Governance*. Oxford: OUP.
Scott, R. W. (2008). *Institutions and organizations—Ideas and interests*. London: Sage Publications.
Sheehan, M. (2007). *The international politics of space*. London: Routledge.
Smith, L. J., & Hörl, K. U. (2007). Institutional challenges for space activities in Europe. *Acta Astronautica, 60,* 210–220.
Steinmo, S., Thelen, K., & Longstreth, F. (1992). *Structuring politics—Historical institutionalism in comparative analysis*. Cambridge: Cambridge University Press.
Streeck, W., & Thelen, K. (2005). *Beyond continuity. Institutional change in advanced political economies*. Oxford: OUP.
Suzuki, K. (2003). *Policy logics and institutions of European space collaboration*. Burlington: Ashgate.

Tallberg, J. (2003). *European governance and supranational institutions—Making states comply*. London & New York: Routledge.

Thelen, K. (1999). Historical institutionalism in comparative politics. *Annual Review of Political Sciences, 2,* 369–404.

Thornton, P. H., Jones, C., & Kury, K. (2005). Institutional logics and institutional change in organizations: Transformation in accounting, architecture, and publishing. *Transformation in Cultural Industries Research in the Sociology of Organizations, 23,* 125–170.

Times Higher Education. (2004). New constitution gives green light to EU space activities. *Times Higher Education*, July 14, 2004. Retrieved from: https://www.timeshighereducation.com/news/new-constitution-gives-green-light-to-eu-space-activities/190137.article. August 12, 2016.

Wallace, H. (2000a). The institutional setting. In H. Wallace & W. Wallace (Eds.), *Policy-making in the European Union* (pp. 3–38). Oxford: OUP.

Wallace, H. (2000b). The policy process. In H. Wallace & W. Wallace (Eds.), *Policy-making in the European Union* (pp. 39–64). Oxford: OUP.

Waltz, K. N. (1979). *Theory of international politics*. New York: McGraw-Hill Publishing Company.

Wendt, A. (2001). Driving with the rearview mirror: On the rational science of institutional design. *International Organization, 55*(4), 1019–1049.

Wörner, J. (2012). A heated debate concerning 'a cold potato'. *DLR Blogs—Jan Woerner's Blog*. Retrieved from http://www.dlr.de/blogs/en/archive/jan-woerner/a-heated-debate-concerning-a-cold-potato.aspx. Accessed March 18, 2017.

Wörner, J. (2013). Von der kalten Kartoffel zum toten Pferd. *DLR Blogs—Jan Woerner's Blog*. Retrieved from: http://www.dlr.de/blogs/de/desktopdefault.aspx/tabid-5896/9578_read-632/. Accessed March 18, 2017.

Young, O. R. (2010). Regime dynamics: The rise and fall of international regimes. In J. L. Goldstein & R. H. Steinberg (Eds.), *International institutions—Volume II consequences: When, where and why international institutions are effective* (pp. 163–183). London: Sage Publisher.

Chapter 5
Conclusion and Synthesis

While European Space collaboration was initially developed outside the competences of the EU with space programmes being carried out almost exclusively under the framework of ESA and national agencies, the EU has gained "parallel competences" in space policy following the adoption of the Lisbon Treaty. Currently, the EU and ESA work together under a FA but the EU Commission has recently published a Communication entitled "European Space Policy" (ESP), which gave rise to a renewed debate on the roles of EU and ESA in European space policy. Even though ESA's Member States have agreed to keep ESA as an intergovernmental organisation during the ESA Ministerial Council of 2014, the discussion about ESA becoming part of the EU framework continues. With the EU and ESA both acting in the field of European space activities with differing but substantially "dual memberships" and partially overlapping competencies, questions regarding coherence (Rüttgers, 1989, p. 127), closer cooperation (Hobe, Kunzmann, & Reuter, 2006, p. ix) and the adaptability particularly with regard to the institutional arrangement gave rise to an academic and political debate about the future institutional set-up and division of labour between the EU and ESA (cf. for example Hörber, 2016a, p. 53). In this context, several scenarios for an institutional set up or a division of labour between EU and ESA have been debated both academically and politically.

Against this background, the "hot potato" currently under debate among experts in the space community is: "What will happen to ESA?" Since academics are no fortune tellers, this dissertation cannot anticipate future developments. This dissertation conducted a historic analysis of the development of the inter-institutional relationship between the EU and ESA, guided by the question: **To what extent has ESA changed institutionally so far?** This is in line with Thelen (1999), who claims that knowing how institutions were constructed "provides insights into how they might come apart" (p. 400).

ESA is not the only institution, becoming questioned, as the EU extends its policy field. Rather the evolution of the EU as an international actor, integrating ever more policy fields, led to a process of institutional change of several already existing international organisations. Some of which, were included as EU agencies (the WEU

N.-L. Remuss, *Theorising Institutional Change: The Impact of the European Integration Process on the Development of Space Activities in Europe*, Springer Theses, https://doi.org/10.1007/978-3-319-95978-8_5

Satellite Centre (cf. Davara, 2000), the WEU Security Institute (cf. Council of the European Union, 2006) and the WEAO & WEAG) into the community framework, others were dissolved (the WEU, the JAA), and a third group of institutional actors managed to co-exist (CERN, CEPT). Thus, the case of the inter-institutional relationship between the EU and ESA might give interesting insights into the following question: **"What is the impact of the European integration process on existing institutional actors?". When, where and how are the EU's supranational institutions such as the Commission able to implement their own agenda?**

5.1 Approach

To answer these questions, this dissertation considered the development of space activities in Europe as a case study of institutional change and provided an updated overview of the development of space policy in Europe. This has been called for by Hörber (2016b, p. 254), who underlined that space as a policy domain has not been studied substantially yet. To do so a chronological approach has been chosen. This is in line with Dudley & Richardson (1999), who consider long-term historical policy studies as the only way to understand the development of institutional change (Hörber, 2016b, p. 254). Bearing in mind the importance of the time frame for the analysis of institutional change (cf. Aminzade 1992; Campbell 2004, pp. 44–6), the development of space activities in Europe was divided into six phases. These were separated based on critical events (cf. Chap. 3):

1. Phase 1: The historic evolution of ESA (1959–1974)
2. Phase 2: From the founding of ESA to the entering of the EC in the space domain (1975–1986)
3. Phase 3: From the first steps of the EC to the ESA-EU Framework Agreement (1986–2004)
4. Phase 4: From the Framework Agreement to the Lisbon Treaty (2004–2009)
5. Phase 5: Post-Lisbon Treaty developments (2009–2012)
6. Phase 6: Post-ESA-EU Declaration of November 2012 (Post-2012).

Chapter 2 has complemented this empirical analysis with a review of the state of the art of the literature, dealing with institutional change and the European integration process. It gave an overview of the applicable tenants of European Integration Theories (i.e. Neofunctionalism, Intergovernmentalism and Liberal Intergovernmentalism, Supranational Governance), Agency Theories, Theories of International Relations and New Institutionalism. Concluding that there seems to be a lack of a structured approach to the analysis of institutional change, with different authors using different definitions of the term "institutions" and describing different aspects as institutional change, the development of a more structured approach was the objective of this dissertation. In line with Keohane & Nye (2001, p. 50) the goal was to identify factors, which are most important in describing institutional change. In line with Héritier (2007, p. 1) this dissertation aims to fill the gap in the existing research on institutional transformation once institutions have been put into place. In line with Shepsle (1989a, b, p. 133) it was sought to shed light on the selection, survival, adaptation, and evolution of institutional practices and to identify different types of change as called for by Campbell (2004, p. 31).

Since most scholars agree that no single theory can account adequately for everything (Egeberg, 2004, p. 199), but rather theories offer "a selective grasp on reality" (Héritier, 2007, p. 39), and are valuable "because they structure all observations" (Rosamond, 2000, p. 4), this dissertation used different theories as complementing each other. Such an approach is in line with the second movement in institutional analysis of the 1990s (Campbell, 2004, p. 4). Hence all theoretic accounts under consideration were understood as answering complementary aspects of the same puzzle. Thus, the existing theories and theoretic accounts on institutional change were analysed in terms of what they offer regarding the analysis of institutional change.

5.2 Relevance

Until recently space policy and ESA attracted very little serious and thoughtful attention from politicians or the general public, being regarded as a field of experts and Sci-Fi geeks. This has changed in the past decade with the potential benefits of space exploration being F discovered by politicians and economists alike (Hörber, 2009b, p. 412). Space policy is a particularly interesting case. First, ESA was founded as an independent institution entirely separate from the European integration process that resulted in the EU. Thus, as Hörber (2016d) put it, "the fact that there is now close cooperation with the EU is significant and worthy of investigation" (Hörber, 2016a, p. 54). Secondly, space policy has the potential to foster what European integration has promised: peace and prosperity, while it will always have the potential to attract the mind and hearts of many (Hörber, 2016d, p. xxiii). Directly or indirectly space is relevant to every EU policy area (Hörber, 2016c, p. 11). It is very much at the cutting edge of current EU policy developments. Recent phenomena such as climate change, rapid developments in crisis management and prevention etc. all need space capabilities to "manage" them (Hörber, 2016c, p. 1). As Hörber (2016b) mentioned, space policy itself can also be considered a driving force of European integration. As such it has not yet been sufficiently understood because "space has not been studied as a policy domain as have defence, environment, transport or climate change." (p. 254). As Hörber (2009b) put it, the literature on the European space effort is quite limited. While reference can be found in books dealing mainly with security-related issues, a structured analysis of the European space effort is missing (Hörber, 2009b, p. 406). Thus, while taking the existing literature as a starting point, this study went back to the historic sources of European space collaboration.

At the same time, the question on how to explain both institutional stability and change is a central concern in political science today (Greif & Laitin, 1994, p. 634). Sigalas (2016, p. 77) most recently underlined the need for "a fully-fledged theory of change". The long-term study of the development of space policy in Europe provides for an answer to the calls of these three academics. Thereby, special attention has been paid to the role of the European Commission the process of institutional change, something which Pollack (Pollack 1997b, p. 111) called for. At the same time, Doleys (2000, p. 546) raised the question in what situations governments can be expected to dominate policy-making, and under what supranational actors.

Thus, this study built on two blocks:

- the history of space policy in Europe.
- existing theories (theories of institutional change, empirical observations form the institutional developments as part of the European integration process, insights from theoretic accounts on institutional design).

5.3 Methodology

ESA as an institution and the development of space policy in Europa were considered a case of institutional change. In line with Keohane & Nye (2001, p. 50), the objective was to identify factors, which are most important in describing institutional change. By considering the existing theoretic approaches to institutional change as complementary lenses rather than competing schools, those were identified as "stages of institutional development" and "facets of institutional change", a facet being defined as "one of several parts",[1] being synonymous to "aspects". Thus, this dissertation proposed to structure the analysis of institutional change, by relying on the following:

- the stages of institutional development
  - agenda formation (Levy, Young, & Zürn, 1995),
  - institutional choice (i.e. selection and design) (Ibid.),
  - operationalisation (Ibid.) or evolution (Florensa, 2004),
  - redefinition (Ibid.) and
  - crisis (Ibid.).[2]
- facets of institutional change
  - the degree of change (incremental or revolutionary change; in which dimension it took place),
  - the pace of institutional change (fast and slow changes),
  - sources of institutional change (endogenous, exogenous),
  - change agents and
  - forms of institutional change.

[1]cf. Longman Dictionary of Contemporary English.

[2]In line with this are two aspects to the evolution of an institution: first, the conditions under which institutions come about (agenda formation) and second, the process of competitive selection by relevant actors between different alternating institutional forms (institutional selection, choice or design); cf. Florensa (2004), p. 9–10; Hall and Taylor (1996), p. 945.

Thereby answering the research question "When or why do institutions change?" and "which different types of institutional change exist". The idea was to develop a theoretic framework, which can be applied to different cases, making them comparable. Considering the development of ESA as a case study for institutional change, this dissertation highlights the advantages and disadvantages of such an approach.

5.4 Conclusions

To understand institutional change, the dissertation analysed the case of ESA in terms of stages of institutional development and facets of institutional change. Combining the findings from the analysis of the state of the art of theories, explaining both the European integration process as well as institutional change, with the empirical findings. Also, the six phases of the development of space policy in Europe were analysed in terms of the five facets of institutional change as identified, with the objective

- to provide an updated overview of the development of space policy in Europe
- to advance the debate about the impact of the European integration process on existing institutional actors
- to make a theoretical contribution to our understanding of institutional change
- to make a theoretical contribution to our understanding of when supranational actors are able to advance their own agenda.

This section summarises the advantages, disadvantages, and findings of such an approach to institutional change.

5.4.1 The Degree of Institutional Change of ESA

This dissertation provided an updated overview of the development of space policy in Europe. This is a gradual development of the existing work on the historic evolution of space activities in Europe and one of the few political science perspectives on it. Given much academic and political debate on the role of ESA as part of the space governance structure in Europe, one might ask "To what extend has ESA changed institutionally so far?". By applying the identified facets of institutional change to the case of ESA, it becomes possible for the first time to provide a structured answer to this question. This dissertation, has in line with Hall (1993, p. 278) and Campbell (2004, p. 36, 48), suggested to differentiate between

- simple change, i.e. in instruments and means,
- complex change, i.e. change in the normative (e.g. principles prescribing goals of behaviours and appropriate ways to pursue them) institutional dimension,

- change in the regulative (e.g. legal, constitutional or other rules that constrain and regulate behaviour) dimensions of an institution, and
- change in the cultural cognitive dimension (e.g. culturally shaped assumptions about reality and the frames through which it is perceived, understood and given meaning) (Campbell, 2004, p. 36, 48).

As to qualify the degree of change. Throughout the whole period of institutional development of ESA under discussion, no simple change in instruments and means, complex change or regulative change could be observed. Thus, one could say ESA has, if at all, changed in the cultural cognitive dimension, i.e. in its assumptions about reality and the frames through which it is perceived.

5.4.2 *The Impact of the European Integration Process on the Development of Space Policy in Europe*

The dissertation also sought to advance the debate about the impact of the European integration process on existing institutional actors, thereby making a theoretical contribution to our understanding of when supranational actors are able to advance their own agenda. The answer to this is twofold and will be found below. First, the European integration process can be considered an exogenous source for institutional change (cf. Sect. 5.4.4.3). Secondly, the European Commission as an institution can be considered a change agent (cf. Sect. 5.4.4.4.).

One of the two main exogenous sources for institutional change, which could be identified in the six phases under analysis, was indeed the European integration process. In five out of six phases, the impact of the European integration process was identified as an exogenous source for institutional change and in four out of six phases, the Commission was found to have acted as a change agent. Merely in the second phase, discussions in the context of the wider politics of European integration did not have an impact on the space field. This can be explained by Europe's integration ambitions coming to a halt in the 1960s and 1970s within the EC, when the Franco German differences caused a stalemate and put a stop to the efforts to achieve a common design for nuclear reactors.

5.4.3 *Stages of Institutional Development as a Way to Structure the Observations on Institutional Change*

One way to structure the observations of institutional change that was proposed by this dissertation was, to rely on Levy et al. (1995) and Florensa's (2004) stages of institutional development. The idea was to specify the time frame over which institutions are tracked, to determine how much institutional change has occurred. This is something which for example has been called for by Campbell (2004, pp. 41–3).

Having applied these stages of institutional development to the case of ESA, the question arises, whether this proofed a fruitful endeavour. Indeed, Levy et al. (1995) and Florensa's (2004) stages of institutional development were perceived as a good starting point and were considered as helpful in guiding the analysis of institutional change. They offer a way to structure the observations.

Moreover, given Sandholtz & Stone Sweet's (1998, p. 16) claim about institutions being in constant evolution, one of the research questions of this dissertation was "Which stages of institutional development can be identified?". The case study under analysis has shown that the identified phases can be complemented by a possible sixth phase of institutional development. Following a period of redefinition is not necessarily a crisis but there can also be a phase of „consolidation". Alternatively, an institution can also be consolidating instead of experiencing a crisis. Moreover, the case study has shown that the phases of „consolidation" and „revision" are dominant. Founding phases and crisis are rather rare. This can also be seen in the table below. One might conclude from this observation that institutions are constantly revising and consolidating. The case study has also shown that in practice, it can be difficult to determine when one of the phases starts and ends. They should thus be considered on a continuum.

Moreover, it was hoped to see whether the reasons for selection of a particular institution, which have been identified by Pollack (1996, p. 434), Wendt (2001, p. 1023–4) and Wallace (2000, pp. 43–4) as rational expectations about costs and benefits (Wendt, 2001, p. 1043) and normative considerations of what is appropriate (Wendt, 2001, pp. 1023–4), can be identified in the case study or whether those institutions were chosen unintentionally as suggested by Heritier (2007, p. 1). Furthermore, it was hoped to discover, in how far the founders of the institutions were constrained by the availability of institutional alternatives (Jupille, 2004, p. 1) and borrowed from prior institutions as suggested by Jupille (2004) and Scott (2008, p. 94). Also it was hoped to discover in how far contextual factors, derived from the broad circumstances, functional factors, derived from some of the core functions of politics and demands to be met, and institutional factors, i.e. opportunities and constraints, which influence in how far institutions are able to channel and to structure the behaviour of political actors from the participating countries (Wallace, 2000, pp. 43–4), were relevant in the process of institutional selection in the case of ESRO, ELDO and ESA. Indeed, ELDO, ESRO and ESA have been selected intentionally based on rational expectations about the costs and benefits, through a process of intergovernmental bargaining on the basis of national preferences and other motivational factors. None of them came about unintentionally. Existing institutions, such as CERN, EURATOM and NATO, inspired the institutional choice of the founders of ESRO and ELDO. ESA benefitted from the experience of ESRO and ELDO and the pitfalls in their institutional design. Thus, Jupille (2004, p. 1) and Scott (2008, p. 94) can be said to be right, as ESRO, ELDO and ESA borrowed from prior existing institutions. ELDO was created based on political will, while ESRO was created given the growing recognition among the scientific community, of the stakes involved in space (Gaubert & Lebeau, 2009, p. 37).

It was also hoped to see whether the conditions for the foundation of an institution, i.e. the "problem condition", the "cognitive condition", the "hegemonic condition" (Rittberger & Zangl, 2006, p. 25), "windows of opportunity", and breaking points, which have been identified on the basis of Rittberger & Zangl (2006, p. 25), DiMaggio (1988, p. 14) and Olsen (1997, pp. 209–10, 217) were present when ESRO, ELDO and ESA were founded. Indeed, the proposed conditions necessary for the foundation of an institution could be identified in the case of ESRO and ELDO. It was not possible to identify the hegemonic condition in the case of ESA.

5.4.4 Facets of Institutional Change as a Way to Structure the Observations on Institutional Change

Some of the research questions posed at the beginning of this dissertation were "What are the sources for institutional change?", "Which factors are most important in describing institutional change?", "Which outcomes of institutional change are possible?" and "How can different cases of institutional change best be compared?".

To answer these, this dissertation has identified five "facets of institutional change" on the basis of the existing literature on institutional change. It was consequently proposed to analyse institutional change in terms of these. A facet was understood as "one of several parts",[3] being synonymous to "aspects".

To make a theoretical contribution to our understanding of institutional change and to structure the debate on this topic, the question arises whether the proposed facets of institutional change proofed to be suited, to structure the debate on institutional change. Some of which might still be too superficial as to qualify institutional change.

5.4.4.1 Degree of Institutional Change

First, the degree of change as a dimension to qualify institutional change proofed rather difficult to distinguish. The differentiation between simple change, i.e. in instruments and means, as well as complex change, i.e. in goals and ends (Hall, 1993, p. 278) proved to be too superficial, as all phases under discussion qualified for simple change. Even when a new institution was newly founded, it had the same goals and ends, and thus no complex change could be identified. At the same time, the more differentiated distinction in changes in the regulative, normative, and cultural cognitive institutional dimension proved insufficient to grasp all facets of the case under discussion. In practice, it seemed to be rather difficult to identify change in the cultural cognitive dimension, which does not seem to be an objective characteristic, but can rather be subjective in nature. Further research has to be conducted as to find a suitable theoretic framework to further differentiate the degree of change.

[3] cf. Longman Dictionary of Contemporary English.

5.4.4.2 Pace of Institutional Change

Secondly, the pace of institutional change, which was distinguished in slow, incremental through small adjustments, gradual or evolutionary in a continuous manner and rapid revolutionary change through large steps in a discontinuous manner, resulting in a "steep transformation", breakdown and replacement of old institutional arrangements (Campbell, 2004, p. 33; Madariaga, 2011, p. 17, 29; Roland, 2004, pp. 116–7), proved to be a sufficient theoretic framework to qualify institutional change. Incremental changes as suggested by Historical Institutionalists [i.e. North 1990, p. 73; March & Olsen 1989, p. 169; Pollack 1996, p. 441; Campbell 2004, p. 48], seem to provide the "main story" (March & Olsen, 1989, p. 169), as four out of six phases under analysis can be considered to have changed in an incremental manner. Only the founding of ESA (phase 1 & phase 2), replacing ELDO and ESRO, can be considered a large step in a discontinuous manner. The case of space policy also highlighted how small incremental changes can culminate in evolutionary change, an example being the incorporation of space policy in the Lisbon Treaty as a shared competence.

5.4.4.3 Sources of Institutional Change

Third, it was proposed to distinguish between endogenous [i.e. Greif & Laitin 1994, p. 634; Streeck & Thelen 2005; Young 2010] and exogenous sources [i.e. Pollack 1996, p. 441; Krasner 1988, p. 67; Scott 2008; Steinmo, Thelen, & Longstreth 1992; Suzuki 2003] of institutional change. Endogenous sources would be, self-undermining processes (Greif & Laitin, 1994, p. 634), institutional design (Héritier, 2007; Streeck & Thelen, 2005, p. 19) or internal contradictions (Young, 2010, p. 174), ideological innovation, learning processes through the acquisition of knowledge and skills and changes in tastes or preferences of the actors involved (Florensa, 2004, p. 6). Exogenous sources of institutional change would be "critical junctures or punctuation" (cf. Historical Institutionalism), i.e. changes in the attribution of natural resources, extreme climatic events or natural disasters,[4] changing preferences of supportive/influencing governments (Pollack, 1996, p. 441), which might result from unfulfilled expectations, shifts in power distributions of previously supportive coalitions or the appearance of new actors (Florensa, 2004, p. 18; Pollack, 1996, p. 438; Scott, 2008, p. 198), changing voter preferences, leading to new political alignments, lobby of business interest (Scott, 2008, p. 196), broad changes in the policy environment of social, political or economic nature (Florensa, 2004, p. 18; Pollack, 1996, p. 438),[5] and improvements in the quality of information about either

---

[4]For more information on "critical junctures" refer to the following literature cf. Florensa (2004), p. 18; Christiansen and Vanhoonacker (2008); Campbell (2004), p. 26; Hall and Taylor (1996), p. 945; Krasner (1984), p. 235 in Ikenberry (1994), p. 9; Horak (2007), p. 21; March and Olsen (2006), p. 12.

[5]For example changes in the markets, technological innovations or legislative changes Steinmo, Thelen, and Longstreth (1992), p. 16.

the operation of the institution or the environment within which it operates (Pollack, 1996, p. 438). It was expected to find both endo- and exogenous sources when analysing the case study. In four out of six phases under analysis both endogenous and exogenous sources of institutional change could be identified. Since the second phase mainly deals with the „operationalisation" and founding of ESA naturally neither endogenous nor exogenous sources for change could be identified. The differentiation between endogenous and exogenous sources of change is difficult during the fourth phase, as this is characterised by the resolutions issued by the Space Council consisting of both the ESA Council and the EU Competitiveness Council. The two main exogenous sources, which were identified in the six phases under analysis, were the impact of the European integration process and of the geopolitical context mainly coming from the U.S. In fact, in five out of six phases under discussion, institutional change can be said to have resulted from the geopolitical context.

While such a distinction proved useful for guiding the analysis and making different cases of institutional change comparable, the question arises whether the mere distinction in exogenous and endogenous sources is fruitful, if both seem to be present in most phases under discussion. Also, there is a hen-egg-problem inherent in the distinction, since in practice, it can be hard to distinguish the original source, be it the self-undermining process (endogenous source) or the broad changes in the policy environment (exogenous source), which might have led to the self-undermining process in the first place. Conditions under which endogenous or exogenous sources preside could not be identified. Additionally, academics did not clearly define whether they considered changing preference of Member States of an intergovernmental institution as exogenous factors or endogenous sources of institutional change. This dissertation decided to consider them as exogenous sources.

Exogenous Source—the U.S.

This dissertation has shown that changes in institutional contexts of socio-economic or political nature can influence existing institutions. The external environment of the "space race"-era shaped and limited the choices of institutional arrangements at European level. Thus, not only gave the U.S., the impetus for Europe to engage in the space field, in order not to be left behind, it was also the policies of the U.S. government and the commercialisation of the space market, that gave rise to the changes in the development of space policy in Europe.

When in 1965 for example the U.S. satellite Early Bird demonstrated for the first time the feasibility of telecommunications by satellite, thereby "opening up an entire new field of immense commercial and military potential" (Krige, 1994a, p. 2), the objectives of the European programme had to be re-evaluated. This lead to the realisation that ELDO's approach was no longer sustainable. In line with Historical Institutionalism, this event can also be interpreted as a critical juncture, leading to a new path—the foundation of ESA. The commercial success of telecommunications programmes by the U.S. government and industry changed the policy environment around the space development (cf. Sect. 4.2.3.2.).

Another example of the U.S. influence on space developments in Europe, can be exemplified by the European wish to be independent and autonomous from the U.S., which in the context of satellite navigation, gave the impetus for the EU to take its role as an actor in the area by providing the framework for the European satellite navigation system Galileo (cf. Sect. 4.2.3.2.).

Exogenous Source—the European Integration Process

One of the two main exogenous sources for institutional change, which could be identified in the six phases under analysis, was the European integration process. In five out of six phases, the impact of the European integration process could be identified as an exogenous source for institutional change. Merely in the second phase, discussions in the context of the wider politics of European integration did not have an impact on the space field, which can be explained by Europe's integration ambitions coming to a halt in the 1960s and 1970s within the EC, when the Franco German differences caused a stalemate and put a stop to the efforts to achieve a common design for nuclear reactors.

The question of the British membership, which played a profound role in the demise of ELDO during the first phase under analysis, can be considered an example of the impact of the European integration process on the development of space activities in Europe. Also, the entrance of the EC/EU into the space area during the second and the third phase, can be considered another example.

5.4.4.4 Change Agents

Fourth, it was suggested that sources of institutional change "have to be activated by agents willing to generate change" (Madariaga, 2011, p. 24), such as States (and their ministries, representatives), individuals, groups (interest groups and communities of experts) or institutions (administrative staff of international organisations; parliamentary assemblies of international organisation) (Rittberger & Zangl, 2006, p. 78). The case under discussion showed an existence of all four groups of change agents across the six phases under analysis, i.e. States (in all six phases), individuals (five out of six phases), groups such as epistemic communities (two out of six phases) and other institutions such as the Commission (in four out of six phases) could be identified to have acted as a change agent.

States

In line with Doleys (2000) one of the research questions was "in what situations governments can be expected to dominate policy-making, and under what conditions supranational actors can be expected to affect change independently of and even in the face of opposition from governments?" It was expected to find out that States acted as change agents activating institutional change in all six phases under discussion, given that space policy is a matter of national sovereignty, a symbol of scientific and technical capability, an index of political power and a component of military strategy, and thus governed by States. Indeed, States were found to have acted as change agents in all six phases under discussion.

Moreover, the current status quo of the inter-institutional relationship between ESA and the EU is a result of intergovernmental bargaining at the ESA Ministerial Council in November 2014. Keeping everything as it is, could be regarded as the "lowest common denominator". Several authors refer to institutions chosen by Member States as "equilibrium institution", meaning that "no possible coalition of actors with an institutional right to change the existing institutions, has an incentive to do so" (Pollack, 1996, p. 437). This is what Scharpf came to refer to as the "Joint-Decision Trap". Accordingly, a given institution, once instituted, tends to remain in place, rigid and inflexible, even in the face of a changing policy environments due to intergovernmentalism and decisions by unanimity (Pollack, 1996, p. 440). Moreover, it is clear, that Europe can only compete with the U.S., Russia and China when working together. However, given that space is a matter of national sovereignty, States do not want to give up their influence in European space collaboration by integrating ESA as a EU agency into the EU. This would result in ESA being subordinated to a DG of the Commission and thus space collaboration would be decided by a non-elected supranational institution. Instead, ESA is currently being governed by the ESA Ministerial Council, composed of the national ministers in charge of space activities of the ESA Member States. Thus, ESA by its very nature is an intergovernmental organisation (Wörner, 2013).

The case of the foundation of ESRO and ELDO further highlights when States influence presides over the one of individuals or experts. While ESRO can be said to be the result of the call of the scientific community and mainly Amaldi's and Auger's power of negotiations, ELDO was the result of intergovernmental negotiations and not scientific considerations. This can be explained by the fact that ELDO dealt with launchers, which can be considered a question of national sovereignty. Thus, one could say that States' influence presides over the influence of individuals (experts) in questions of national sovereignty. To generalise this conclusion, it needs to be tested in different case studies.

Individuals

The case study shows that intergovernmental bargains, as proclaimed by Neo-Realism, were not the only driving force of European space collaboration and institutional change. Scientists, engineers and other experts were also involved in the creation and change of the institutions of European space collaboration. Both Neorealism and Neoliberal institutionalism, which neither account for the role of epistemic communities nor for the role of individuals, like Auger and Amaldi, are able to fully explain the case of European space collaboration. On the contrary, the case shows that individuals can play an important if not essential role in the agenda formation by highlighting the need for cooperation. Fligstein (1997a, 1997b, 2010), North (1990, p. 73), Sabatier & Jenkins-Smith (1996) and Sabatier, (1998, p. 109)[6] have considered so-called "institutional entrepreneurs" as the main source for institutional change, defining them as individuals, who have their own preferences, creating opportunities for innovation and institutional change (Thornton, Jones, & Kury, 2005, p. 129). Their assumptions can be underlined by the case at hand.

In line with Fligstein (1997a, p. 398), one area of particular interest of this dissertation was to analyse, whether in times of uncertainty or when highly technical issues are being discussed, skilful individuals with high social capital and the ability to access decision makers are able to advance their own agenda and act as change agents or can take the lead in founding institutions. Since space policy is of highly technical nature, it was expected that individuals were able to act as change agents in all phases under discussion. Individuals were found to have acted as change agents in five out of six phases under discussion. Only during the foundation process of ESA, no influence of individuals could be identified. Auger and Amaldi were considered as credible in the policy field, given their experience in founding CERN. Their social capital as well as their success in framing the story about the need for European space collaboration can be attributed to their past in CERN. In line with this, Campbell (2004) has highlighted the role of individuals as change agents in the context of bricolage; i.e. institutional change involving the recombination of old institutional elements (Campbell, 2004, p. 28; Lanzara, 1998, p. 26). In line with Campbell (2004), it is not surprising that Auger and Amaldi, given their experience with CERN, recombined institutional elements in innovative ways, leading to the institution building and the foundation of ESRO.

Groups

Another area of particular interest of this dissertation was, whether in times of uncertainty or when highly technical issues are being discussed, policy-makers' demand information, which are provided by expert groups such as epistemic communities (Haas, 1992, p. 12, 2010). In two out of six phases, individuals who were part of an epistemic community, acted as change agents and are thus considered as groups acting as change agents. As has been shown, the scientific community acted as one of the driving sources for the foundation of ESRO and during the third phase, when the Commission tasked the Wise Men to provide independent advice for promoting the evolution of the ESA-EU relationship and the necessary institutional change. Both are examples of groups' influence in the context of institutional change.

[6] cf. Dudley and Richardson (1999), p. 227.

Institutions and Institutional Interrelations

According to theoretic accounts such as Neofunctionalism, Supranational Governance (cf. Hix, Pierson), and Rational Choice Institutionalism, supranational institutions exert significant independent influence on institutional and policy outcomes (Hix, 2005, p. 16), act as driving forces (Pierson, 1998, p. 29), play an autonomous role (Andersen, Eliassen, & Sitter, 2001, p. 28) and can act as change agents.

Since the influence of supranational actors is not a new phenomenon, the interesting question is, when are supranational actors able to advance their own agenda, which was thus one of the research questions of this dissertation. It was expected that the Commission would be able to advance its own agenda regarding topics were it has a great authority, legitimacy, or "institutional standing" and "institutional embeddedness", when it has the control of information and knowledge, when it is able to abuse information asymmetries, when it is able to frame or shape a proposal in such a way, that it is securing approval even in the light of Member State(s)'s opposition, or when it acts together with powerful allies such as States. None of which could be empirically proven.

Nevertheless, in four out of six phases,[7] the Commission, an institution, can be identified to have acted as a change agent. This is striking, as the Commission was a new actor to the field of space policy and thus lacked authority, legitimacy, or institutional standing in this policy field. States were also not particularly fragmented, when the Commission entered the policy field. Moreover, being of interest to national sovereignty, the space field was expected to be prone to States' influence and not supranational actor's influence. The Commission's influence can partly be explained by enjoying several windows of opportunities such as the Kyoto Protocol and the Balkan conflicts (cf. Sects. 3.3.3 and 4.2.3.2) (Marta & Stephenson, 2016, p. 99). The Commission was able to use this to frame space policy in a way, that it is a tool to reach the Community's objectives, thus justifying an extended role of the Commission in this policy field. Another explanation could be the Commission's right of initiative inherent in the Community legislation.

5.4.4.5 Forms of Institutional Change

Fifth, while theoretically it seems to be relatively easy to distinguish between the five different outcomes of institutional change, as suggested by the literature under discussion, i.e. salience/ dissolution, different services/ends, institutional adjustment/

[7] In the first phase, it was the European integration process as a contextual factor influencing the field of space policy and the development of ESA rather than the institution Commission. In the second phase, the founding phase of ESA, no influence of the Commission could be identified, which can be explained by the Franco German differences to the common design for nuclear reactors, which caused a stalemate and made Europe's integration ambitions coming to a halt in the 1960s and 1970s. Since the fourth phase under analysis, a central role of the Commission in advancing the debate about defining the roles of the different players, especially EU and ESA in the space arena, and in identifying a governance model for the space domain could be identified.

learning, bricolage and translation, the case study showed, that in practice it is relatively more complicated to categorise institutional change according to these options. While the first and the second phase seem to cover all possible outcomes, apart from the institution being put in the service of different ends, the other phases can all be characterised as institutional learning and adaptation.

5.5 Assessment, Implications and Generalisation

The objectives of this study were fourfold:

- to provide an updated overview of the development of space policy in Europe
- to advance the debate about the impact of the European integration process on existing institutional actors
- to make a theoretical contribution to our understanding of when supranational actors are able to advance their own agenda
- to make a theoretical contribution to our understanding of institutional change.

An assessment of these four objectives will follow below.

First, while several authors have perceived ESA and the inter-institutional relationship or space governance structure in Europe worth the analysis, the literature on the European space effort is still quite limited. While references can be found in books dealing mainly with security-related issues, a structured analysis of the European space effort was missing. Thus, while taking the existing literature as a starting point, this study went back to the theoretic accounts on institutional change and the historic sources of European space collaboration to provide an up-to-date analysis of the development of space policy in Europe from a political science perspective. To do so, a chronological approach has been applied. To be able to structure the observations, the development of space activities in Europe was divided into six phases. These were separated based on critical events. Alternatively, it would also be possible to divide the history of space policy in Europe in just four phases, according to the actors involved, being: phase 1: The historic evolution of ESA (1959–1974); phase 2: From the founding of ESA until the Lisbon Treaty (1975–2011); phase 3: Post-Lisbon Treaty developments (2011–2014); phase 4: The New Space era. Another alternative, would be to divide the history of space developments in Europe in phases according to applications, i.e. Space 1.0: Astronomy, Space 2.0: Space Race (Apollo-Era), Space 3.0: Applications and initial Cooperation (i.e. ISS) and Space 4.0: Infrastructure, world-wide cooperation and commercialisation. Nonetheless, a division in six phases proved useful for the area of interest of this dissertation. It had the advantage of providing a more detailed analysis, given the division into more phases. This shows that the choice of the time frame is dependent on the subject of analysis.

By applying the stages of institutional development and facets of institutional change to the six phases of the historic development of space policy in Europe, it was sought to answer the research question "In how far has ESA changed institutionally so far?", coming to the conclusion that ESA has, if at all, changed in the cultural cognitive dimension, i.e. in its assumptions about reality and the frames through which it is perceived (cf. Sect. 5.4.1).

Secondly, the debate about the impact of the European integration process on existing institutional actors was advanced by considering the European integration process as an exogenous source of institutional change and the EU's Commission as a change agent.

Third, to make a theoretical contribution to our understanding of when supranational actors are able to advance their own agenda, it was observed when the Commission was able to advance its own agenda. It was concluded that supranational actors do not need to have authority, legitimacy, or institutional standing in the policy field to advance its own agenda but can exploit windows of opportunities. In the case of the Commission, it can as well make use of its right of initiative and justify its action by declaring the policy field as a tool to reach the Community's objectives.

Fourth, by considering the existing theoretic approaches to institutional change as complementary lenses rather than competing schools, this dissertation has taken a step towards a more structured approach to the analysis of institutional change. This consists of a combination of stages of institutional development and facets of institutional change, thereby seeking to answer the research questions "When or why do institutions change?" "What are the sources of institutional change?", "Which stages of institutional development can be identified?", "Which different types of change exist?", "Which factors are most important in describing institutional change?" and "How can different cases of institutional change best be compared?". It made a theoretical contribution to our understanding of institutional change, by analysing the existing literature on institutional change in terms of what it offers regarding the degree and pace of institutional change and the sources for institutional change. Additionally, different actors were identified as change agents. Considering the development of ESA as a case study for institutional change, this dissertation highlighted the advantages of such an approach (cf. Sects. 5.4.3 and 5.4.4), which can consequently be applied to more case studies. All facets of institutional change have been identified during all phases of the development of space policy in Europe. This is in line with Richardson's observation that the complexity of the EU policy process requires multiple models and the utilisation of "concepts from a range of models to at least describe the policy process accurately" (Richardson, 1996, p. 48). Thus, the case study proves that different theories should indeed be considered as complementary rather than competing, since they answer complementary aspects of the same puzzle.

5.6 Further Research

This dissertation is limited to the period of analysis of European space collaboration from 1959 until 2014 and can thus only provide a snap shot of the policy field. Since the end of the period of observation in 2014, new actors such as private entrepreneurs have entered the space arena. Consequently, research has to be continued.

By focusing on the role of the Commission, the findings of this dissertation regarding the impact of the European integration process on existing institutional actors are limited. Further research is needed, considering the role of the Council and the European Parliament as to gain a complete picture of the impact of the European integration process on existing institutional actors. Additionally, ESA is not the only case, where the evolution of the EU as an international actor, integrating ever more policy fields, led to a process of institutional change of already existing international organisations. Other actors were included into the community framework as EU agencies (the WEU Satellite Centre (cf. Davara, 2000), the WEU Security Institute (cf. Council of the European Union, 2006) and the WEAO & WEAG), were dissolved (the WEU (Bailes, 2011), the JAA), or managed to co-exist (CERN, CEPT). Against this background the question arises, how to account for the different institutional outcomes resulting from the impact of the European integration process. Being limited to the case of ESA, this dissertation cannot answer this question. To answer this question more case studies have to be conducted and compared.

While both, the stages of institutional development and the facets of institutional change, were considered as helpful in guiding the analysis of institutional change and were perceived to provide a good starting point, they should be further tested and developed by applying them to more cases. Also, further research has to be conducted to find a suitable theoretic framework to qualify the degree of change as explained above.

The example of ELDO and ESRO also gives insights into the question of institutional stability. While ESRO proved to be successful and was transferred and renamed as ESA, ELDO was struggling right from the start. One possible explanation to this discrepancy, could be that ELDO was created based on political will, while ESRO was created given the growing recognition among certain communities, primarily the scientific community, of the stakes involved in space (Gaubert & Lebeau, 2009, p. 37). Further research into the question if institutions based on political will are stronger than those based on the will of e.g. scientific communities is needed.

5.7 Outlook

The discussions on the inter-institutional relationship between the EU and ESA are said to have lost its topicality in 2014. Observers argue that the window of opportunity opened by the discussions on the Multiannual Financial Framework of the EU, which was agreed upon on 27th June 2013, is now closed and today's discussions remain largely characterised by vanities.

Assuming that the role of the EU in space affairs will continue to grow, the influence of the EU institutions over the development of the European space policy will also increase (Sigalas, 2016, p. 255). Coupled with the ongoing debate over the general governance of space activities in Europe and the ability of the Commission to propose itself as an actor, it is logical to assume that the ambitions of the EU, and the Commission, in space will continue to grow and not be limited to the current programmes or existing frame sets (Marta & Stephenson, 2016, p. 112). “Dealing with it by saying that the ESA must become an agency of the Community would be like imagining the problem can be “magicked” away” (Gaubert & Lebeau, 2009, p. 43) is however a too simple approach. “The crux of the problem consists of establishing relations between two entities of profoundly different character so that the ESA can become the executive arm of Brussels in space matters without losing its own dynamism.” (Ibid.) It will therefore need a lot of good will and political courage on both sides as well as a common vision about the challenges to make substantial progress under the present framework conditions (Hörber, 2016d, p. xxiii). In line with Gaubert & Lebeau (2009) the problem is not solved by creating a melting pot of EU and ESA but by using them to the “benefit of European construction while preserving their identity and dynamism” (p. 43). Also, we need to be realistic about the time frame required for such an endeavour. From a purely institutional point of view, space policy is an EU responsibility. Europeanisation of space policy and space activities is thus often considered[8] the next logical step. The EU however lacks sufficient funds and expertise in leading concrete projects. This expertise is concentrated in ESA, which is intergovernmental in its structure and goes beyond the EU, when it comes to membership. ESA’s activities are based on the principle juste retour, which is not applied in the context of EU integration. As has been shown ESA was deliberately designed as a non-political but intergovernmental institution. It is often even argued that the reason ESA has been successful is precisely because it is not under the EU framework and that “the EU administration is already bloated and would therefore be unable to manage a space programme properly” (Hörber, 2016a, p. 59).

Moreover, space has recently entered the so-called “New Space Era” with commercial companies such as Space-X and Blue Origin entering the market, leading to increasing competition and national space budgets getting under pressure. They have aspired to a wide range of activities including communications, remote sensing, mining, 3D-printing and even tourism (Aliberti & Krasner, 2016, p. 147). In satellite-based applications the roles of industry, commercial operators and new entrepreneurs have developed tremendously over the last fifteen years, resulting in the introduction of new concepts, industry-initiated ventures, development of new or emerging markets, technology initiatives and public-private partnerships (Brachet, 2016, p. 212). With the increasing commercialisation of the space arena, pressure on a European governance[9] structure will increase, given questions related to regulation of commercial space activities, policy formulation, security and international relations in the light of changing roles of traditional space actors such as national space agencies (Ibid., p. 214). At the same time the number of State engaging in space activities

[8]cf. For example Hörber, T. “Foreword.” European Space Policy. Eds. T. Hörber & P. Stephenson. London & New York: Routledge (2016). xx–xxiii. xxii.

[9]The task of governance is to define a policy and put it into practice. cf. Gaubert and Lebeau (2009), p. 37.

will further increase, with economic interests such as competitiveness and growth and aspects of international cooperation being the focus (Giannopapa, Adriaensen, & Sagath, 2016, p. 189). In this context, theoretic frameworks on institutional change will be further pressurised as to account for the new private actors in space, what may lead to a new branch of theory, focusing on post-statehood developments.

Space policy is not an important factor in winning public support or national election campaigns. However, space policy has the potential to foster what European integration has promised: peace and prosperity (Hörber, 2016d, p. xxii). Innovation remains at the heart of global competitiveness, ensuring peace and security (Ibid., p. xxi). Mastering space is a precondition for the survival and defence of the European way of life (Ibid.) as it is a tool for reaching many of the EU policy objectives. European space policy (and ESA?) has outgrown its merely technical roots acquiring political relevance (Hörber, 2016a, p. 62). A European Space Policy has the potential to become a component of Europe's "soft power" expertise to be deployed among other elements as an effective tool of foreign policy. It may even be a considerable element towards strengthening the European identity (Hörber, 2016a, pp. 62–3; Rathgeber & Remuss, 2009; Remuss, 2011). International space activities foster a climate of political cooperation and help in responding to Europe's political priorities such as climate change, sustainable development, humanitarian aid, and security. Europe's space activities illustrate its role and position in the international system (La Regina, 2016, p. 203). It is thus often said, that a stable equilibrium between ESA and the EU as a precondition for a European Space Policy to develop, will be decisive for European integration to reach its full potential (Hörber, 2016a, p. 63).

Taking a last look at the analysis at hand, ESA with its large capital stock of expertise seems to be a prime example of institutional stability regardless the continuous pushes coming from the EU. It remains to be seen, whether we will just experience another example of the famous quote "plus ça change, plus c'est la meme chose" (Ebbinghaus, 2005, p. 17).

References

Aliberti, M., & Krasner, S. D. (2016). Governance in space. In C. Al-Ekabi, B. Baranes, P. Hulsroj, & A. Lahcen (Eds.), *Yearbook on space policy 2014* (pp. 143–166). Vienna: Springer.

Aminzade, R. (1992). Historical sociology and time. *Sociological Methods & Research, 20,* 456–480.

Andersen, S. S., Eliassen, K. A., & Sitter, N. (2001). Formal processes: EU institutions and actors. In S. S. Andersen & K. A. Eliassen (Eds.), *Making policy in Europe* (pp. 20–43). London: SAGE Publisher.

Bailes, A. J. K. (2011). Death of an Institution. The end for Western European Union, a future for European defence? *Egmont Paper 46*. Retrieved from: http://aei.pitt.edu/32322/1/ep46.pdf. Accessed 13 Mar 2017.

Brachet, G. (2016). The optimum role of governments in space. In C. Al-Ekabi, B. Baranes, P. Hulsroj, & A. Lahcen (Eds.), *Yearbook on space policy 2014* (pp. 211–215). Vienna: Springer.

Campbell, J. (2004). *Institutional change and globalization*. Princeton & Oxford: Princeton University Press.

Christiansen, T., & Vanhoonacker, S. (2008). At a critical juncture? Change and continuity in the institutional development of the council secretariat. *West European Politics, 31*(4), 751–770.

Council of the European Union. (2006). *Council joint action amending joint action 2001/554/CFSP on the establishment of a European Union Institute for security studies (2006/1002/CFSP)*. Brussels: European Communities.

Davara, F. (2000). The future of the WEU Torrejon centre. *Air & Space Europe, 2*(4), 24–25.

DiMaggio, P. (1988). Interest and agency in institutional theory. In L. G. Zucker (Ed.), *Institutional patterns and organizations: Culture and environment* (pp. 3–23). Cambridge: Ballinger Pub.

Doleys, T. (2000). Member states and the European Commission: Theoretical Insights from the new economics of organization. *Journal of European Public Policy, 7*(4), 532–553.

Dudley, G., & Richardson, J. (1999). Competing advocacy coalitions and the process of „frame reflection": A longitudinal analysis of EU steel policy. *Journal of European Public Policy, 6*(2), 225–248.

Ebbinghaus, B. (2005). *Can path dependence explain institutional change—Two approaches applied to welfare state reform*. MPIfG Discussion Paper 05/2. Retrieved from: http://www.mpifg.de/pu/mpifg_dp/dp05-2.pdf. Accessed 17 Mar 2017.

Egeberg, M. (2004). An organisational approach to European integration: Outline of a complementary perspective. *European Journal of Political Research, 43,* 199–219.

Fligstein, N. (1997a). Social skill and institutional theory. *American Behavioral Scientist, 40*(4), 397–405.

Fligstein, N. (1997b). Social skill and the theory of fields. Retrieved from: http://www.irle.berkeley.edu/culture/papers/Fligstein01_01.pdf. Accessed 13 Mar 2017.

Fligstein, N. (2010). Institutional entrepreneurs and cultural frames—The case of the European Union's single market program. *European Societies, 3*(3), 261–287.

Florensa, M. C. (2004). *Institutional Stability and Change. A Logic Sequence for Studying Institutional Dynamics*. Paper presented at the Tenth Biennial Conference of the International Association for the Study of Common Property (IASCP). Oaxaca, Mexico. Retrieved from: https://dlc.dlib.indiana.edu/dlc/bitstream/handle/10535/1823/CostejaFlorensa_Institutional_040525_Paper220.pdf?sequence=1. Accessed 13 Mar 2017.

Gaubert, A., & Lebeau, A. (2009). Reforming European space governance. *Space Policy, 25,* 37–44.

Giannopapa, C., Adriaensen, M., & Sagath, D. (2016). The member states of the European space agency—National governance structures, priorities and motivations for engaging in space. In T. Hörber & P. Stephenson (Eds.), *European space policy* (pp. 173–190). London & New York: Routledge.

Greif, A., & Laitin, D. D. (1994). A theory of endogenous institutional change. *American Political Science Review, 98*(4), 633–652.

Haas, P. M. (1992). Introduction: Epistemic communities and international policy coordination. *International Organization, 46*(1), 1–35.

Haas, P. M. (2010). Introduction: Epistemic communities and international policy coordination. In J. L. Goldstein & R. H. Steinberg (Eds.), *International institutions—Volume I* (pp. 185–220). London: Sage.

Hall, P. A. (1993). Policy paradigms, social learning, and the state: The case of economic policymaking in Britain. *Comparative Politics, 25*(3), 275–296.

Hall, P. A., & Taylor, R. C. R. (1996). Political science and the three new institutionalisms. *Political Studies, XLIV,* 936–957.

Héritier, A. (2007). *Explaining institutional change in Europe*. New York: OUP.

Hix, S. (2005). *The political system of the European Union*. Houndmills: Palgrave.

Hobe, S., Kunzmann, K., & Reuter, T. (2006). *Forschungsbericht ESA—EU: Rechtliche Rahmenbedingungen einer zukünftigen kohärenten Struktur der europäischen Raumfahrt*. Kölner Schriften zum Internationalen und Europäischen Recht Bd. 13. Berlin: LIT Verlag.

Horak, M. (2007). *Governing the post-communist city: Institutions and democratic development in Prague*. Toronto: University of Toronto.

Hörber, T. (2009). The European space agency and the European Union: The next step on the road to the stars. *Journal of Contemporary European Research, 5*(3), 405–414.
Hörber, T. (2016a). The European space agency and the European Union. In T. Hörber & P. Stephenson (Eds.), *European space policy* (pp. 53–65). London & New York: Routledge.
Hörber, T. (2016b). Conclusion: „to boldly go where no one has gone before". In T. Hörber & P. Stephenson (Eds.), *European space policy* (pp. 253–263). London & New York: Routledge.
Hörber, T. (2016c). Introduction—Towards a European space policy. In T. Hörber & P. Stephenson (Eds.), *European space policy* (pp. 1–12). London & New York: Routledge.
Hörber, T. (2016d). Foreword. In T. Hörber & P. Stephenson (Eds.), *European space policy* (pp. xx–xxiii). London & New York: Routledge.
Ikenberry, G. J. (1994). *History's heavy hand: Institutions and the politics of the state*. Paper prepared for a conference on "New Perspectives on Institutions", University of Maryland. Retrieved from: https://scholar.princeton.edu/sites/default/files/HistorysHeavyHand_0.pdf. Accessed 13 Mar 2017.
Jupille, J. (2004). *Procedural politics—Issues, influence, and institutional choice in the European Union*. Cambridge: Cambridge University Press.
Keohane, R., & Nye, J. (2001). *Power and interdependence*. New York: Longman.
Krasner, S. (1988). Sovereignty—An institutional perspective. *Comparative Political Studies, 21*(1), 66–94.
Krige, J. (1994a). The European space system. In J. Krige & A. Russo (Eds.), *Reflections on Europe in space* (pp. 1–13). ESA HSR-11. Nordwijk: ESA.
La Regina, V. (2016). Space as a field and tool of international relations. In T. Hörber & P. Stephenson (Eds.), *European space policy* (pp. 191–207). London & New York: Routledge.
Lanzara, G. F. (1998). Self-destructive processes in institution building and some modest countervailing mechanisms. *European Journal of Political Research, 33,* 1–39.
Levy, M. A., Young, O. R., & Zürn, M. (1995). The study of international regimes. *European Journal of International Relations, 1,* 267–330.
Madariaga, A. (2011). *Patterns of institutional change and external competitiveness in neoliberal and dependent political economies. The cases of chile and estonia*. Budapest: Central European University.
March, J. G., & Olsen, J. P. (1989). *Rediscovering institutions: The organizational basis of politics*. New York: The Free Press.
March, J. G., & Olsen, J. P. (2006). Elaborating the 'new institutionalism'. In R. A. W. Rhodes, S. A. Binder, & B. A. Rockman (Eds.), *The oxford handbook of political institutions* (pp. 3–20). Oxford: OUP.
Marta, L., & Stephenson, P. (2016). Role of the European commission in framing European space policy. In T. Hörber & P. Stephenson (Eds.), *European space policy* (pp. 98–113). London & New York: Routledge.
North, D. C. (1990). *Institutions, institutional change and economic performance*. Cambridge: Cambridge University Press.
Olsen, J. P. (1997). Institutional design in democratic context. *The Journal of Political Philosophy, 5*(3), 203–229.
Pierson, P. (1998). The path to European integration: A historical institutionalist perspective. In W. Sandholtz & A. Stone Sweet (Eds.), *European integration and supranational governance* (pp. 27–58). Oxford: OUP.
Pollack, M. A. (1996). The new institutionalism and EC governance: The promise and limits of institutional analysis. *Governance: An International Journal of Policy, Administration, and Institutions, 9*(4), 429–458.
Pollack, M. A. (1997). The commission as an agent. In N. Nugent (Ed.), *At the heart of the union—Studies of the European Commission* (pp. 111–130). Houndmills: Macmillan.
Rathgeber, W., & Remuss, N. (2009). *Space security—A formative role and principled identity for Europe*. ESPI Report 16. Wien: ESPI, 2009.

Remuss, N. (2011). Meeting Europe's external and internal security through space applications—Security as an element for European identity. In B. Baranes & K.-U. Schrogl (Eds.), *European Identity through Space* (pp. 132–144). Vienna: Springer.
Richardson, J. (1996). Actor-based models of national and EU policy making. In H. Kassim & A. Menon (Eds.), *The European Union and national industrial policy* (pp. 26–51). London: Routledge.
Rittberger, V., & Zangl, B. (2006). *International organization—Polity, politics and policies*. New York: Palgrave Macmillan.
Roland, G. (2004). Understanding institutional change: Fast-moving and slow-moving institutions. *Studies in Comparative International Development, 38*(4), 109–131.
Rosamond, B. (2000). *Theories of European integration*. Houndmills: Palgrave.
Rüttgers, J. (1989). *Europas Wege in den Weltraum: Programme—Proteste—Prognosen*. Frankfurt am Main: Umschau Verlag.
Sabatier, P. A. (1998). The advocacy coalition framework: Revisions and relevance for Europe. *Journal of European Public Policy, 5*(1), 98–130.
Sabatier, P. A., & Jenkins-Smith, H. S. (1996). *The advocacy coalition framework: An assessment*. Paper presented to the Annual Meeting of the American Political Science Association.
Sandholtz, W., & Stone Sweet, A. (1998). *European Integration and Supranational Governance*. Oxford: OUP.
Scott, R. W. (2008). *Institutions and organizations—Ideas and interests*. London: Sage Publications.
Shepsle, K. A. (1989a). Institutional equilibrium and equilibrium institutions. *Journal of Theoretical Politics, 1*(2), 131–147.
Shepsle, K. A. (1989b). Studying institutions: Some lessons from the rational choice approach. *Journal of Theoretical Politics, 1*(2), 121–147.
Sigalas, E. (2016). Europe in space—The European parliament's justification arsenal. In T. Hörber & P. Stephenson (Eds.), *European space policy* (pp. 66–81). London & New York: Routledge.
Steinmo, S., Thelen, K., & Longstreth, F. (1992). *Structuring politics—Historical institutionalism in comparative analysis*. Cambridge: Cambridge University Press.
Streeck, W., & Thelen, K. (2005). *Beyond continuity. Institutional change in advanced political economies*. Oxford: OUP.
Suzuki, K. (2003). *Policy logics and institutions of European space collaboration*. Burlington: Ashgate.
Thelen, K. (1999). Historical institutionalism in comparative politics. *Annual Review of Political Sciences, 2*, 369–404.
Thornton, P. H., Jones, C., & Kury, K. (2005). Institutional logics and institutional change in organizations: Transformation in accounting, architecture, and publishing. *Transformation in Cultural Industries Research in the Sociology of Organizations, 23*, 125–170.
Wallace, H. (2000). The policy process. In H. Wallace & W. Wallace (Eds.), *Policy-making in the European Union* (pp. 39–64). Oxford: OUP.
Wendt, A. (2001). Driving with the rearview mirror: On the rational science of institutional design. *International Organization, 55*(4), 1019–1049.
Wörner, J. (2013). Von der kalten Kartoffel zum toten Pferd … DLR Blogs—Jan Woerner's Blog. Retrieved from: http://www.dlr.de/blogs/de/desktopdefault.aspx/tabid-5896/9578_read-632/. Accessed 18 Mar 2017.
Young, O. R. (2010). Regime dynamics: The rise and fall of international regimes. In J. L. Goldstein & R. H. Steinberg (Eds.), *International institutions—Volume II consequences: When, where and why international institutions are effective* (pp. 163–183). London: Sage Publisher.

About the Author

Nina-Louisa Remuss is Deputy Head of the Berlin Office of the German Aerospace Center (DLR). Before that she assisted a German Member of Parliament at the German Bundestag in Berlin (Germany) and worked as an Associate Fellow and Project Manager for the European Space Policy Institute (ESPI) in Vienna (Austria). There, she published and presented several studies and articles on space and security. She holds a Master's Degree (with distinction) in International Security Studies from the University of St. Andrews (United Kingdom) and a Bachelor's Degree in European Studies from the University of Maastricht (The Netherlands) and spent an exchange semester at the University of Bologna (Italy). Since 2011 she started her Ph.D. studies at the University of Potsdam (Germany), alongside working fulltime for the German Aerospace Center (DLR).

N.-L. Remuss, *Theorising Institutional Change: The Impact of the European Integration Process on the Development of Space Activities in Europe*, Springer Theses, https://doi.org/10.1007/978-3-319-95978-8

Bibliography

Andersen, S. S., & Eliassen, K. A. (2001a). *Making policy in Europe*. London: Sage.

Andersen, S. S., & Eliassen, K. A. (2001b). Informal processes: Lobbying, actor strategies, coalitions and dependencies. In S. S. Andersen & K. A. Eliassen (Eds.), *Making policy in Europe* (pp. 44–60). London: Sage.

Ansell, S. (2006). Network institutionalism. In R. A. W. Rhodes, S. A. Binder, & B. A. Rockman (Eds.), *The oxford handbook of political institutions* (pp. 75–89). Oxford: OUP.

Bache, I., & George, S. (2006). *Politics in the European Union*. Oxford: OUP.

Balme, R., & Chabanet, D. (2008). *European governance and democracy: Power and protest in the EU*. Plymouth: Rowman & Littlefield Publishers Inc.

Becker, F., & Worms, J. C. (1996). Defining a European space research policy—The role of the European space science committee. *Space Policy, 12*(4), 277–286.

Blyth, M. (2002). *Great transformations: The rise and decline of embedded liberalism*. Cambridge: Cambridge University Press.

Boix, C., & Stokes, S. C. (2007). *The oxford handbook of comparative politics*. New York: OUP.

Borras, S. (2007). The European commission as network broker. *European Integration Online Papers, 11*, 1.

Burzykowska, A. (2006). ESDP and the space sector—Defining the architecture and mechanisms for effective cooperation. *Space Policy, 22*, 35–41.

Bush, P. D. (1987). The theory of institutional change. *Journal of Economic Issues, XXI.3*, 1075–1116.

Campbell, J. (2007). Institutional reproduction and change. First draft. Forthcoming In G. Morgan, J. L. Campbell, C. Crouch, P. H. Kristensen, O. K. Pedersen & R. Whitley (Eds.), *The oxford handbook of comparative institutional analysis*. New York: OUP. Retrieved from: http://citeseerx.ist.psu.edu/viewdoc/download?doi=10.1.1.465.1572&rep=rep1&type=pdf. Accessed 17 Apr 2017.

Checkel, J. T. (2005). International institutions and socialization in Europe: Introduction and framework. *International Organization, 59*, 801–826.

Commission of the European Communities. (1992). *The European community and space: Challenges, opportunities and new actions* (COM (92) 360 final). Retrieved from http://aei.pitt.edu/5806/1/5806.pdf. Accessed 13 Mar 2017.

Commission of the European Communities. (2004). *Green paper—European space policy (COM (2003) 17 final)*. Brussels: European Communities.

Commission of the European Communities. (2011). *Towards a space strategy for the European Union that benefits its citizens (COM (2011) 152 final)*. Brussels: European Communities.

N.-L. Remuss, *Theorising Institutional Change: The Impact of the European Integration Process on the Development of Space Activities in Europe*, Springer Theses, https://doi.org/10.1007/978-3-319-95978-8

Commission of the European Communities. (2016). *Space strategy for Europe* (COM(2016)705 final). Retrieved from: http://ec.europa.eu/docsroom/documents/19442. Accessed 13 Mar 2017.

Costa, O., & Jorgensen, K. A. (2013). The influence of international organisations on the European Union: A political science perspective. In R. A. Wessel & S. Blockmans (Eds.), *Between autonomy and dependence—The EU legal order under the influence of international organisations* (pp. 69–84). The Hague: Aser Press.

Council of the European Union. (2003). *Resolution on the development of an overall European space policy (2003/C 149/05)*. Brussels: European Communities.

Cox, R. (1986). Social forces, states and world orders: Beyond international relations theory. *Journal of International Studies, 10*(2), 126–155.

Crawford, S. E. S., & Ostrom, E. (1995). A grammar of institutions. *The American Political Science Review, 89*(3), 582–600.

Danesy, D. (Ed.). (2007). Agenda 2011: A document by the ESA director general and the ESA directors. Nordwijk: ESA. Retrieved from: http://www.esa.int/esapub/br/br268/br268.pdf. Accessed on 25 Mar 2017.

Dehousse, R., & Thompson, A. (2012). Intergovernmentalist in the commission: Foxes in the henhouse? *European Integration, 34*(2), 113–132.

Eberlein, B. (2004). New governance in the European Union: A theoretical perspective. *Journal of Common Market Studies, 42*(1), 121–141.

Elgström, O., & Jönsson, C. (2000). Negotiation in the European Union: Bargaining or problem-solving? *Journal of European Public Policy, 7*(5), 684–704.

Elgström, O., & Smith, M. (2000). Negotiation and policy-making in the European Union—Processes, systems and order. *Journal of European Public Policy, 7*(5), 673–683.

ESA. (2001). *Council joint action on the establishment of the European Union Satellite Centre (EUSC) (*2001/555/CFSP). OJ L 200. Retrieved from: http://www.europarl.europa.eu/meetdocs/2009_2014/documents/sede/dv/sede021213counciljointaction_/sede021213counciljointaction_en.pdf. Accessed 19 Mar 2017.

Fligstein, N., & McNichol, J. (1998). The institutional terrain of the European Union. In W. Sandholtz & A. Stone Sweet (Eds.), *European integration and supranational governance* (pp. 59–91). Oxford: OUP.

Froehlich, A. (2010). Space and the complexity of European rules and policies: The common project Galileo and GMES—Precedence for a new European legal approach? *Acta Astronautica, 66,* 1262–1265.

Garrett, G. (1992). Internal cooperation and institutional choice: The European community's internal market. *International Organization, 46*(2), 533–560.

Gaubert, A. (2006). Is there really any duplication in Europe's space activities? *Space Policy, 22,* 1–2.

Gellner, W., & Strohmeier, G. (Eds.). (2005). *Politische Strukturen und Prozesse im Wandel.* Baden-Baden: Nomos.

Göhler, G. (1996). Institutions in political theory: Lessons for European integration. In D. Rometsch & W. Wessels (Eds.), *The European Union and member states* (pp. 1–19). Manchester: Manchester University Press.

Goldstein, J. L., & Keohane, R. O. (1993). *Ideas and foreign policy—Beliefs, institutions and political change*. New York: Cornell University Press.

Grant, W. (1993). Pressure groups and the European community: An overview. In S. Mazey & J. Richardson (Eds.), *Lobbying in the European community* (pp. 27–46). Oxford: OUP.

Greenwood, J., & Ronit, K. (1994). Interest groups in the European community: Newly emerging dynamics and forms. *West European Politics, 17*(1), 31–52.

Hall, P. A., & Taylor, R. C. R. (1998). *The potential of historical institutionalism: A response to Hay and Wincott* (pp. 958–962). XLVI: Political Studies.

Hartlapp, M., Metz, J., & Rauh, C. (2010a). *How external interests enter the European Commission: Mechanisms at play in legislative position formation*. Retrieved from: http://bibliothek.wzb.eu/pdf/2010/iv10-501.pdf. Accessed 30 Mar 2017.

Hartlapp, M., Metz, J., & Rauh, C. (2010b). The agenda set by the EU Commission: The result of balanced or biased aggregation of positions? *LEQS Paper, 21*. Retrieved from http://www.lse.ac.uk/europeanInstitute/LEQS%20Discussion%20Paper%20Series/LEQSPaper21.pdf. Accessed 30 Mar 2017.

Hartlapp, M., Metz, J., & Rauh, C. (2010c). *Made in Brussels—Wie externe Interessen ihren Weg in die Politikformulierung der EU Kommission finden*. Retrieved from: http://www.ssoar.info/ssoar/handle/document/30901. Accessed 30 Mar 2017.

Hay, C. (2006). Constructivist institutionalism. In R. A. W. Rhodes, S. A. Binder, & B. A. Rockman (Eds.), *The oxford handbook of political institutions* (pp. 56–73). Oxford: OUP.

Hay, C., & Wincott, D. (1998). Structure, agency and historical institutionalism. *Political Studies, XLVI*, 951–957.

Hayward, K. (2010). Europe in space. *The RUSI Journal, 155*(6), 36–42.

Hayward, J., & Menon, A. (2003). *Governing Europe*. Oxford: OUP.

Hill, C., & Smith, M. (2005). *International relations and the European Union*. Oxford: OUP.

Hinnfors, J. (1999). Stability through change: The pervasiveness of political ideas. *Journal of Public Policy, 19*(3), 293–312.

Hix, S. (1998). The study of the European Union II: The "new governance" agenda and its rival. *European Journal of Public Policy, 5*(1), 38–65.

Hofmann, H. C. H., & Türk, A. H. (2006). *EU administrative governance*. Cheltenham: Edward Elgar.

Hull, R. (1993). Lobbying Brussels: A view from within. In S. Mazey & J. Richardson (Eds.), *Lobbying in the European Union* (pp. 82–92). Oxford: OUP.

Hurrell, A., & Menon, A. (1996). Politics like any other? Comparative politics, international relations and the study of the EU. *West European Politics, 19*(2), 366–402.

Hwang, H., & Powell, W. W. (2007). Institutions and entrepreneurship. In H. Hwang & W. W. Powell (Eds.), *The handbook of entrepreneurship—Handbook of entrepreneurship research* (pp. 179–210). Dordrecht/Boston/London: Kluwer Publishers.

Immergut, E. M. (1998). The theoretical core of new institutionalism. *Politics Society, 26*(5), 5–34.

Immergut, E. M., & Andersen, K. M. (2008). Historical institutionalism and West European politics. *West European Politics, 31*(1-2), 345–369.

Jachtenfuchs, M. (1995). Theoretical perspective on European governance. *European Law Journal, 1*(2), 115–133.

Jachtenfuchs, M., & Kohler-Koch, B. (1996). *Europäische integration*. Opladen: Leske & Budrich.

Johnson, R. E. (2007). Europe's space policies and their relevance to ESDP. *Military Technology, 31*(3), 24–37.

Juncos, A. E., & Pomorska, K. (2006). Playing the Brussels game: Strategic socialisation in the CFSP council working groups. *European integration online papers, 10*. Retrieved from: http://eiop.or.at/eiop/pdf/2006-011.pdf. Accessed 2 Apr 2017.

Kaltenecker, H. (1974). Zur Gründung der Europäischen Weltraumorganisation. *Zeitschrift für Luftrecht und Weltraumrechtsfragen, 13*, 244–251.

Kassim, H. (2003). The European administration: Between Europeanization and domestication. In J. Hayward & A. Menon (Eds.), *Governing Europe* (pp. 139–161). Oxford: OUP.

Keohane, R. (1945). *International institutions and state power—Essays in international relations theory*. Boulder/San Francisco/London: Westview Press.

Keohane, R. (1986). *Neorealism and its critics*. New York: Columbia University Press.

Keohane, R., & Martin, L. (1995). The promise of institutionalist theory. *International Security, 20*(1), 39–51.

Kerr, P. (2002). Saved from extinction: Evolutionary theorising, politics and the state. *British Journal of Politics and International Relations, 4*(2), 330–358.

Kerremans, B. (1996). Do institutions make a difference? Non-institutionalism, neo-institutionalism, and the logic of common decision-making in the European Union. *Governance, 9*(2), 217–240.

Kipke, R. (1988). *Die Reaktivierung der Westeuropäischen Union: Zur Perspektive europäischer Sicherheitspolitik*. Münster: lit Verlag.

Knill, C. (2001). *The Europeanisation of national administration—Patterns of institutional change and persistence*. Cambridge: Cambridge University Press.

Kogelnik, M. (2004). *Decentralized organizations on a European level. A definition and new classification analysis of European agencies (EAs) in their role as new regulatory institutions*. Thesis for the M.A.S. of European Studies. University of Vienna.

Kratochwil, F., & Ruggie, J. G. (2010). International organization: A state of the art on an art of the state. In J. L. Goldstein & R. H. Steinberg (Eds.), *International institutions* (Vol. I, pp. 103–123)., Causes London: SAGE Publisher.

Krige, J. (1993). *Pre-history of ELDO—Part I: 1961–1964*. ESA HSR 10. Nordwijk: ESA.

Krige, J., & Russo, A. (1994). *Reflections on Europe in space*. ESA-HSR 11. Nordwijk: ESA.

Lavenex, S. (2004). EU external governance in "Wider Europe". *Journal of European Public Policy, 11*(4), 680–700.

Lehmbruch, G. (1991). The organization of society, administrative strategies, and policy networks. Element of a developmental theory of interest systems. In R. Czada & A. Windhoff-Héritier (Eds.), *Political choice: Institutions, rules, and the limits of rationality* (pp. 121–158). Frankfurt am Main: Westview Press.

Lenzi, G. (1998). The WEU between NATO and EU. *ZEI discussion paper*. Retrieved from: http://aei.pitt.edu/344/. Accessed 2 Apr 2017.

Lepsius, R. M. (1990). *Interessen, ideen und institutionen*. Opladen: Westdeutscher Verlag.

Levinthal, D. A., & March, J. G. (1993). The Myopia of learning. *Strategic Management Journal, 14,* 95–112.

Lewis, O., & Steinmo, S. (2010). Taking evolution seriously: Institutional analysis and evolutionary theory. Retrieved from: http://www.svensteinmo.com/articles/Lewis_Steinmo_2010_taking_evolution.pdf. Accessed 2 Apr 2017.

Lindberg, L. N. (1963). *The political dynamics of European economic integration*. London: OUP.

Lindberg, L. N. (1970). Integration as a source of stress on the European community system. *International Organization, 20*(2), 233–265.

Lindberg, L. N., & Scheingold, S. A. (1970). *Europe's would-be polity—Patterns of change in the European community*. Englewood Cliffs: Prentice-Hall Inc.

Madders, K., & Thiebaut, W. M. (2007). Carpe Diem: Europe must make a genuine space policy now. *Space Policy, 23,* 7–12.

Mahoney, J. (2000). Path dependence in historical sociology. *Theory and Society, 29,* 507–548.

Majone, G. (1997). The agency model: The growth of regulation and regulatory institutions in the European Union. *EIPASCOPE, 3,* 1–6.

Majone, G. (2006). Managing Europeanization: The European agencies. In J. Peterson & M. Shackleton (Eds.), *The institutions of the European Union* (pp. 190–209). Oxford: OUP.

March, J. G., & Olsen, J. P. (1981). Footnotes to organizational change. *Administrative Science Quarterly, 26*(4), 563–577.

March, J. G., & Olsen, J. P. (1991). Exploration and exploitation in organizational learning. *Organization Science, 2*(1), 71–87.

Marks, G., Scharpf, F. W., Schmitter, P. C., & Streeck, W. (1998). *Governance in the European Union*. London: SAGE.

Marsh, D., & Stoker, G. (1995). *Theory and methods in political science*. Houndmills: Macmillian Press Ltd.

Martin, L. L., & Simmons, B. A. (1998). Theories and empirical studies of international institutions. *International Organization, 52*(4), 729–757.

Massey, Sir Harrie. (1986). *History of British space science*. Cambridge: Cambridge University Press.

Mayntz, R. (2004). Governance theory als fortentwickelte Steuerungstheorie? *MPIfG working paper, 4.1*. Retrieved from: http://www.mpifg.de/pu/workpap/wp04-1/wp04-1.html. Accessed 2 Apr 2017.
Mazey, S., & Richardson, J. (1993). *Lobbying in the European community*. Oxford: OUP.
Meyer, M. (2004). *Prinzipale, Agenten und ökonomische Methoden. Von einseitiger Steuerung zu wechselseitiger Abstimmung*. Tübingen: Mohr Siebeck.
Michaelis, A. R. (1988). European space agency. *Interdisciplinary Science Reviews, 13*, 97–191.
Moravcsik, A. (1993a). Introduction: Integrating international and domestic theories of international bargaining. In P. Evans, H. Jacobson & R. Putnam (Eds.), *Double-edged diplomacy: Interactive games in international affairs* (pp. 3–43). Berkeley: University of California Press.
Moravcsik, A. (1993b). Introduction. In P. Evans, H. Jacobson, & R. Putnam (Eds.), *Double-edged diplomacy: Interactive games in international affairs* (pp. 3–42). Berkeley: University of California Press.
Moravcsik, A. (1998). *Centralization or fragmentation—Europe facing the challenges of deepening, diversity and democracy*. New York: Council on Foreign Relations Press.
Moser, P., & Schneider, G. (1997). Rational choice and the governance structure of the European Union: An introduction. *Aussenwirtschaft, 52*(1/2), 11–26.
Müller, M. (2007). The impact of the European integration on institutional change in Switzerland. In H. G. Schröter (Ed.), *The European enterprise: Historical investigation into a future species* (pp. 97–110). Berlin: Springer.
Naja, G. (2001). A joint European strategy for space. *Space Policy, 17*, 83–85.
Nugent, N. (2006). *The government and politics of the European Union*. Houdmills: Palgrave.
Olsen, J. P. (2002). The many faces of Europeanization. *Journal of Common Market Studies, 40* (5), 921–952.
Onoda, M. (2008). Satellite observation of greenhouse gases: Monitoring the climate change regime. *Space Policy, 24*, 190–198.
Orren, K., & Skowronek, S. (1994). Beyond the iconography of order: Notes for a "new institutionalism". In L. C. Dodd & C. Jillson (Eds.), *The dynamics of American politics—Approaches and interpretations* (pp. 311–332). Boulder: Westview Press.
Ostrom, E. (1991). Rational choice theory and institutional analysis: Toward complementarity. *American Political Science Review, 85*(1), 239–243.
Page, E. (1997). *People who run Europe*. Oxford: Clarendon Press.
Peterson, J., & Shackleton, M. (2006). *The Institutions of the European Union*. Oxford: OUP.
Pfaff, D., & Zweifel, P. (1998). Die Principal-Agent Theorie: Ein fruchtbarer Beitrag der Wirtschaftstheorie zur Praxis. *WiSt, 4*, 184–190.
Pierson, P. (1994). The path to European integration: A historical institutionalist perspective. Program for the study of Germany and Europe, working paper No. 5.2. Retrieved from: http://aei.pitt.edu/63633/1/PSGE_WP5_2.pdf. Accessed 29 Mar 2017.
Pierson, P. (2000). The limits of design: Explaining institutional origins and change. *Governance, 13*(4), 475–499.
Reillon, V. (2017). *European space policy—Historical perspective, specific aspects and key challenges*. European Parliamentary Research Service. Retrieved from: http://www.europarl.europa.eu/RegData/etudes/IDAN/2017/595917/EPRS_IDA(2017)595917_EN.pdf. Accessed 17 Apr 2017.
Rhodes, R. A. W. (1995). The institutional approach. In D. Marsh & G. Stoker (Eds.), *Theory and methods in political science* (pp. 42–57). Houndmills: Macmillan.
Rometsch, D., & Wessels, W. (Eds.). (1996). *The European Union and member states. Towards institutional fusion?*. Manchester/New York: Manchester University Press.
Ruggie, J. G. (1975). International responses to technology: Concepts and trends. *International Organization, 29*(3), 557–583.

Ruggie, J. G. (1986). Continuity and transformation in world polity: Toward a neorealist synthesis. In R. Keohane (Ed.), *Neorealism and its critics* (pp. 121–157). New York: Columbia University Press.

Ruggie, J. G. (1993). Territoriality and beyond: Problematizing modernity in international relations. *International Organization, 47*(1), 139–174.

Ruggie, J. G. (2010). International regimes, transactions, and change: Embedded liberalism in the postwar economic order. In J. L. Goldstein & R. H. Steinberg (Eds.), *International institutions—Volume II consequences: When, where and why international institutions are effective* (pp. 199–236). London: Sage Publisher.

Sagan, S. (1994). The perils of proliferation: Organization theory, deterrence theory, and the spread of nuclear weapons. *International Security, 18*(4), 66–107.

Scharpf, F. W. (1988). The joint-decision trap: Lessons from German Federalism and European integration. *Public Administration, 66,* 239–278.

Scharpf, F. W. (1994). Community and autonomy multilevel-policy-making in the European Union. *EUI working paper RSC, 94.1.*

Schimmelfennig, F., & Wagner, W. (2004). Preface: External governance in the European Union. *Journal of European Public Policy, 11*(4), 657–660.

Schmidt-Tedd, B. (2001). Rechtliche Implikationen der gemeinsamen ESA/EU-Raumfahrtstrategie. *Zeitschrift für Luft- und Weltraumrecht, 50*(1), 202–216.

Schmitter, P. C. (1969). Three neo-funcationalist hypotheses about international integration. *International Organization, 23*(1), 161–166.

Smith, M. (2004). Toward a theory of EU foreign policy-making: Multilevel governance, domestic politics, and national adaptation to Europe's common foreign and security policy. *Journal of European Public Policy, 11*(4), 740–758.

Stein, A. A. (2008). Neoliberal institutionalism. In C. Reus-Smit & D. Snidal (Eds.), *The oxford handbook of international relations* (pp. 201–221). Oxford: OUP.

Stevens, A., & Stevens, H. (2001). *Brussels Bureaucrats? The administration of the European Union*. Houndmills: Palgrave.

Stinchcombe, A. (1987). *Constructing social theory*. Chicago/London: The University of Chicago Press.

Stone, A. (1994). What is a supranational constitution? An essay in international relations theory. *The Review of Politics, 56*(3), 441–474.

Streeck, W. (1991). Interest heterogeneity and organizing capacity—Two class logics of collective action. In R. Czada & A. Windhoff-Héritier (Eds.), *Political choice: Institutions, rules and the limits of rationality* (pp. 160–198). Frankfurt am Main: Westview Press.

Trondal, J. (2007). The public administration turn in integration research. *Journal of European Public Policy, 14*(6), 960–972.

Trondal, J., van den Berg, C., & Suvarierol, S. (2008). The compound machinery of government: The case of seconded officials in the European Commission. *Governance, 21*(2), 235–274.

Tsebelis, G., & Garrett, G. (1996). Agenda setting power, power indices and decision-making in the European Union. *International Review of Law and Economics, 16,* 345–361.

Venet, C. (2012). European space governance: The outlook. *Space Policy, 28*(1), 59–60.

Wallace, H. (1992). *The dynamics of European integration*. London/New York: Pinter Publishers.

Wallace, H. (1996). Die Dynamik des EU-Institutionsgefüges. In M. Jachtenfuchs & B. Kohler-Koch (Eds.), *Europäische Integration* (pp. 141–164). Opladen: Leske + Budrich.

Wallace, H. (2000). Analysing and explaining policy. In H. Wallace & W. Wallace (Eds.), *Policy-making in the European Union* (pp. 65–84). Oxford: OUP.

Wallace, H., & Wallace, W. (2000). *Policy-making in the European Union*. Oxford: OUP.

Waltz, K. N. (1986). Political strcutures. In R. Keohane (Ed.), *Neorealism and its critics* (pp. 70–97). New York: Cambridge University Press.

Weaver, R. K., & Rockman, B. A. (1993). *Do institutions matter? Government capabilities in the United States and Abroad*. The Brookings Institutions: Washington, D.C.

Wessel, W. (1996). Institutions of the EU System: Models of explanation. In D. Rometsch & W. Wessels (Eds.), *The European Union and member states* (pp. 20–36). Manchester/New York: Manchester University Press.

WEU. (1991). Decision setting up a WEU Satellite Centre. Retrieved from: http://www.cvce.eu/de/obj/decision_of_the_weu_council_of_ministers_setting_up_a_weu_satellite_centre_vianden_27_june_1991-en-e5018f3a-04fd-43ea-8c4b-6368fe2852dc.html. Accessed 11 Apr 2017.

WEU. (1994). The future WEU Satellite Centre in Torrejon (Doc. 1437).

WEU. (1999). The WEU Satellite Centre—the way ahead. report submitted on behalf of the technological and aerospace committee. Retrieved from: www.cvce.eu/content/publication/2009/10/28/9fb61ce5-6054-4234-84aa-52b8d9601427/publishable_en.pdf. Accessed 29 Jan 2014.

WEU. (2000). *The EU 666 recommendation—The implementation of the common european security and defence policy and WEU's future role—Reply to the annual report of the council*. Paris: WEU.

Wörner, J. (2010). What does the space sector expect from European politics? The challenges ahead for the European space sector and European space policy. Presentation, workshop EISC 2010, Kloster Eberbach, 2 May 2010.

MIX
Papier aus verantwortungsvollen Quellen
Paper from responsible sources
FSC® C105338

If you have any concerns about our products,
you can contact us on
ProductSafety@springernature.com

In case Publisher is established outside the EU,
the EU authorized representative is:
Springer Nature Customer Service Center GmbH
Europaplatz 3, 69115 Heidelberg, Germany

Printed by Libri Plureos GmbH
in Hamburg, Germany